AF556157

Earthworm Ecology and Environment

Satyendra M. Singh
Reader
Department of Animal Science
Mahatma Jyotiba Phule Rohilkhand University
BAREILLY - 243 006, Uttar Pradesh, India

International Book Distributing Co.
(Publishing Division)

Published by

INTERNATIONAL BOOK DISTRIBUTING CO.
(Publishing Division)
Khushnuma Complex Basement
7, Meerabai Marg (Behind Jawahar Bhawan)
Lucknow 226 001 U.P. (INDIA)
Tel. : 91-522-2209542, 2209543, 2209544, 2209545
Fax : 0522-4045308
E-Mail : ibdco@airtelmail.in

First Edition 2009

ISBN 81-8189-062-0

Composed & Designed at :
Panacea Computers
3rd Floor, Agrawal Sabha Bhawan
Subhash Mohal, Sadar Cantt. Lucknow-226 002
Phone : 2483312, 9335927082, 9452295008
E-mail : prasgupt@rediffmail.com

Printed at:
Salasar Imaging Systems
C-7/5, Lawrence Road Industrial Area
Delhi - 110 035
Tel. : 011-27185653, 9810064311

Om Prakash
Vice-Chancellor

Mahatma Jyotiba Phule
Rohilkhand University
Bareilly-243006 (U.P.)

Foreword

Albert Einstein once contrasted the use of power in crossing the seas, in flying across the skies, in sending messages and news over the entire world and in relieving humanity from muscular work with its abuse in twisting and turning the production and distribution system in such a way that everybody must live in fear of being eliminated from the economic cycle and suffering for want of everything and the need of cross country killings at irregular time intervals which make those thinking about future live in fear and terror. The cause of this contradiction, according to him, is the fact that the intelligence and character of the masses are incomparably lower than the intelligence and character of the few who produce something valuable for the community.

The present volume contains contribution of ten such men of incomparably high intelligence and character who produced something valuable for the community in a relatively isolated academic environment of a University or an Institute where creativity is perhaps still possible. They addressed the questions of census, variety, population dynamics and genetics of earthworm, on the one hand, and those of their role in bringing about the second green revolution, degradable solid waste management, interactive contribution of earthworms and Tasar silk worm to forest eco-system and production of silk and self-employment generation in cities and villages.

Credit goes to the editor of the present book Dr. S.M. Singh to involve a large number of individuals from among the farmers in the Extension Programme of the Mahatma Jyotiba Phule Rohilkhand University for taking the scientific methods of improving in an eco-friendly way (through earthworm composting) the soil productivity declining fast due to reckless use of non-eco-friendly chemical fertilizers. The peasant community needs to be fired with the unavoidable need to produce or purchase vermicompost to recoup the rapidly evaporating fertility of soil of their fields. City people as well as unemployed city youth in particular must learn degradable solid waste management from the Extension Centre of the University in order to be pioneers of making the drive for earthworms scavenging for production of vermicompost and sale of earthworms for further proliferation of the movement.

Agrarian development should no longer depend in WTO's Green Room packages imposed from above with the ulterior motive for exploiting the vast Indian markets inbuilt into them. It should respond to the initiative from Indian Universities and create a drive from below, which is helpful in cultivating self-sufficiency.

I have great pleasure in introducing the present book to the world of scholars. I very much wish that the Hindi version of the bok should also come out for the benefit of at least that section of the Hindi speaking farmer communities which is presently at a relatively higher level of intelligence and character just for giving them a little upward pull.

Let not everything be allowed to degenerate into propositions for enormous and overnight commercial gain and self-promotion, not to speak of pervasive brokerage that devour the already paltry means of the masses and close the door for the entry of even a few of them into the circle of men of incomparably high intelligence and character. The education sold to them is hopelessly substandard and made attractive primarily by mid-day meals and "vote for note" scholarship and fee-reimbursements putting their children more on crutches than on their own legs.

(Om Prakash)

Vice-Chancellor
Mahatma Jyotiba Phule Rohilkhand University
Bareilly-243006 (U.P.)

Preface

Since times immemorial, usefulness of earthworms has been emphasized. Burrowing habit of these creatures improves the physical condition of the soil. Aristotle (384-322 B.C.) - one of the greatest philosophers of Greece, called them "intestines of earth". Charles R. Darwin (1809-1882) who was the first naturalist to bring earthworms to the attention of scientific community, nearly one hundred thirty years ago, has stated: "The plough is one of the most ancient and most valuable of man's inventions, but long before he existed, the land was infact regularly ploughed and still continues to be thus ploughed by earthworms". He, in his book, The Formation of Vegetable Mould through the Action of Worms (1881) expressed the opinion that, "earthworms have played a most important part in the history of the world".

However, it was only in the last 35 years that the interest in the research into the ecology and biology of earthworms moved at a fast pace and most of the work on earthworms was compiled in the books, **Biology and Ecology of Earthworms** (Edwards and Lofty, 1972, 1977; Edwards and Bohlen, 1996); **Earthworm Ecology from Darwin to Vermiculture** (Satchell, 1983); **Earthworms: Their Ecology and Relationships with Soils and Land Use** (Lee, 1985) and **Earthworm Ecology** (Edwards, 1998).

In recent years, interest in earthworm diversity and their role in solid waste management have been increasing at an extremely rapid rate and so has interest into the subject. **Earthworms for Solid Waste Management** (Singh, 2007) is a newly published book with fourteen such research articles written by distinguished earthwork ecologists of the country.

The present book, **Earthworm Ecology and Environment,** owes its genesis to the **1st National Symposium on Earthworm Ecology and Environment** held at Mahatma Jyotiba Phule Rohilkhand University, Bareilly (Uttar Pradesh) in April 2007. At this Symposium, attended by nearly 110 scientists from all over the country, 80 research presentations were made in eight different sessions. Out of the total presentations, 18 invited research papers have been selected and edited to form chapters of the book. There are four sections in the book covering all the major aspects of earthworm biology, including earthworm taxonomy, diversity and their multifarious usages in maintaining soil fertility, increasing crop productivity, improving human health and saving environment from solid waste contaminations. These contributions by distinguished earthworm biologists of the country represent the state of the art summary of the ecology and importance of earthworms in Indian context and should be highly valuable not only to the soil scientists alone but also to the environmentalists, ecologists, policy makers ,farmers and students in long term maintenance of soil structure, fertility and pollu tion abatement.

Satyendra M. Singh
Mahatma Jyotiba Phule Rohilkhand University,
Bareilly Uttar Pradesh (India)

Contents

IV. Earthworms in Waste Management

PART - I

EARTHWORM TAXONOMY, DIVERSITY AND POPULATION

Chapter 1

Density, Diversity and Population Dynamics of Earthworm Species in Bareilly Region of U.P.State

Satyendra M.Singh, Om Prakash and Geeta R. Gangwar
Vermiculture and Environmental Research Laboratory, Department of Animal Science, M. J. P. Rohilkhand University, Bareilly –243006 (U.P.)

Soil is the natural habitat for most of the earthworms. They feed upon partly decomposed organic matter as well as on seeds, rooting leaves, ova, larvae and small animals, living or dead, along with the surrounding soil and egest them in the form of 'castings'- rich in macro and micro-nutrients and other components useful to plant growth; while some of the exotic epigeic earthworms like, *Eisenia fetida* and *Eudrilus eugeniae* are very good transformers of various types of solid organic wastes into bio fertilizer. However, few indigenous species of earthworms like, *Perionyx excavatus* and *Perionyx sansibaricus* are known to play a significant role in converting organic waste into vermicompost - a bio fertilizer to augment and sustain fertility status of the soil. Such group of earthworms is useful in reducing the level of environmental pollution and making the nature more natural.

Quantification of density and diversity of earthworm species is an important aspect to understand their abundance in different agro-ecosystems and assess their diversified activities. A number of earthworm species have been reported from various parts of India from different agro-ecosystems (Michaelson, 1909; Gates, 1972; Julka, 1993, 1996; Julka and Paliwal, 1986). Julka (1988) has given a taxonomic monograph of earthworms with 128 species and 26 genera including 6 new genera and 16 new species.

The first record of earthworms in the Indian subcontinent was provided by Robert Templeton (1844); while Stephenson (1923) carried out a detailed survey of earthworms in India including Ceylon (Sri Lanka) and Burma. In the past few years, survey of earthworm population has been carried out by different workers in different states of the country (Chaudhuri and Bhattacharjee, 1999, of Tripura; Julka and Senapati, 1987, of Orissa; Ismail, 1986, of Tamilnadu (Madras); Kale and Krishnamoorthy, 1978, of Karnataka (Bangalore). Recently, biodiversity of earthworms in and around Gwalior region of Madhya Pradesh state has been carried out by Agrawal and Agrawal (2008). Assessment of earthworm population of Uttar Pradesh state has not yet been investigated. Keeping in view such importance of earthworms, a study has been carried out to quantify their density, diversity and population dynamics in Bareilly region of Uttar Pradesh state. The attempt has also been made to identify the earthworm species, which might be used in improving nutrients dynamics of the solid waste.

Study sites, collection and segregation of earthworms

The study was conducted in the three agro-ecosystems (viz. orchard, agricultural and grasslands) of all the four districts (Pilibhit, Shahjahanpur, Budaun and Bareilly) of the region. Nine random samples were taken from all the three areas of each district, using metallic quadrate (size 25 x 25 x 20 cm) with fine cutting edges. Earthworms were collected by hand sorting techniques (Edwards and Lofty, 1977) and segregated into three different size groups (<4 cm-juvenile, ≥ 4cm–immature, and ≥ 8-cm body length-adult (clitellate) and these were identified by observing their external and internal morphological body characters and as per the records of Stephenson (1923) (Table 1 & 2, respectively). Atmospheric and soil temperatures measured by thermometer, organic matter by Walkley and Black (1947) method, pH using systronics pH meter and percent humidity by hygrometer.

Experimental findings

Density of earthworms

Grasslands of the region showed the maximum population density of worms/m^2 area and agricultural lands the minimum out of the selected agro-ecosystems in general **(Fig. 1a.)**. Further, it was noticed that the grasslands and agricultural lands of Pilibhit district have the

Table: 1 – External morphological characters of collected species of earthworms from the region

Earthworm species / Morph. Char.	*Metaphire posthuma*	*Megascolex mauritii*	*Perionyx excavatus*	*Eutyphoeus waltoni*	*Eutyphoeus paivai*	*Eutyphoeus gigas*	*Eutyphoeus pharpingianus*
1. Length (mm)	60-140	80-210	60-120	90-230	150-200	220-250	130-140
2. Diameter (mm)	4-8	3½-5	2-4	4-8	4-6	5-9	4-5
3. Colour	Rich brown.	Dark-yellow &purplish.	Deep purple to reddish & pale.	Brownish to violet-gray & yellowish.	Red-violet brown & grayish.	Purplish brown & pale.	Gray
4. No. of Segments	91-134	166-190	140-146	140-210	120-125	220-236	118-140
5. Prostomium	Tanylobous	Prolobbous/ epilobous	Epilobous	Tanylobous/pro-tanylobous	Tanylobous	Zygolobous	Tanylobous
6. First dorsal pore	12/13	10/11 or 11/12	4/5 or 5/6	12/13 or 11/12	12/13	11/12	11/12
7. Spermathecal pores	5/6-8/9(4 pairs)	6/7-8/9(3 pairs)	7/8-8/9 (2 pairs)	7/8 (1 pair)	7/8 (1 pair)	7/8 (1 pair)	7/8 (1 pair)
8. Clitellum	xiv-xvi=3seg.	xiv-xvii=4 seg.	xiii-xvii=4 seg.	xiii-xvii=5 seg.	xiv-xv (? ½xiii-xvii)	½xiii-½xvii=4 seg.	xiii-xvii(=?).
9. Female pore	xiv (single)	xiv (double)	xiv (single)	xiv (single)	xiv (single)	xiv (single)	xiv seg.
10. Male pore	xviii (paired)	xviii (paired)	xviii (paired)	xviii (paired)	xviii (paired)	xvi-xviii (paired)	xvii (=? Xviii)
11. Genital Markings	xvii and xix	absent	absent	15/16& 18/19 often on 14/15,16/17,rarely on 19/20,20/21.	xix, xx, xxi, xxii, xxiii, xxiv & xxv sometimes on 15/16,16/17 &18/19-22/23.	15/16 (=?).	xiii-xiv or on furrows 13/14-16/17.
12. Calciferous Glands	absent	absent	absent	present (usual).	present (usual).	present (usual), covered 5 seg.	usual.

Table: 2 -Internal morphological characters of collected species of earthworms from the region

Earthworm species / Morph. Char.	*Metaphire posthuma*	*Megascolex mauritii*	*Perionyx excavatus*	*Eutyphoeus waltoni*	*Eutyphoeus paivai*	*Eutyphoeus gigas*	*Eutyphoeus pharpingianus*
1. Gizzard	v seg.	v (=? vi/vii-viii seg.	vi, sometimes on v seg.	vii seg.	xi seg.	viii-ix seg.	vi-vii seg.
2. Septa	5/6-8/9 & 9/10.	7/8-12/13.	No septa thickened	Usual	5/6-7/8.	Usual	Usual(7/8-12/13 much thickened.
3. Spermathecal ampulla	Ovoid	Elongated & constricted	Ovoid	Thick sac-like	Sac-shaped & constricted	Irregular in shape elongated, sac-like	Indistinct/very short.
4. Last pair of hearts	xiv seg.	xiii (=? xiv) seg.	xii seg.	xiv seg.	xiv (=? xiii) seg.	xiii(=?xiv) seg.	xiv seg.
5. Prostatic glands	xvi-xxi seg	xviii-xix seg	xvii-xviii seg	xvii-xix seg.	xvii-xxi (=?xvi-xx) seg.	xi-xiii seg.	Covered 4 seg.
6. Testis	xviii & xix.	x&xi.	x& xi.	xii.	xi-xii.	12/13-13/14.	xi & xii.
7. Seminal vesicles	x, xi & xii seg.	ix & xii seg.	xi & xii-xiv seg.	xii seg.	xvi seg.	10/11 & 12/13/14 groove.	xxxiii seg.
8. Ovary	xx,seg.	xiii(=?xiv) seg.	xiii seg.	xiii seg.	xiii seg.	xiii-xiv_seg.	xiv_seg.
9. Beginning of Intestine	xv seg.	xv seg.	xv seg.	xiv(=?xv) seg.	xv seg.	xv seg.	xiv(=?xv) seg.

maximum earthworm population; while it was the maximum in orchard lands of Bareilly. The lowest density of earthworms was recorded in grasslands, agricultural lands and orchard lands of Badaun, Bareilly and Shahjahanpur districts of the region, respectively.

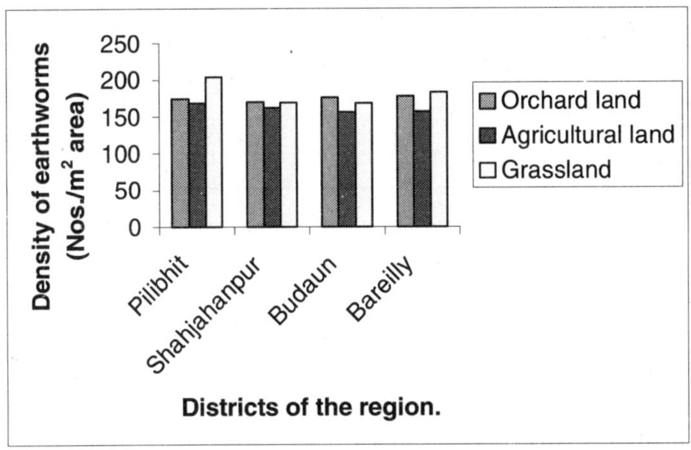

Fig.1a: Total Population density of earthworms in three agro-ecosystems of Bareilly region.

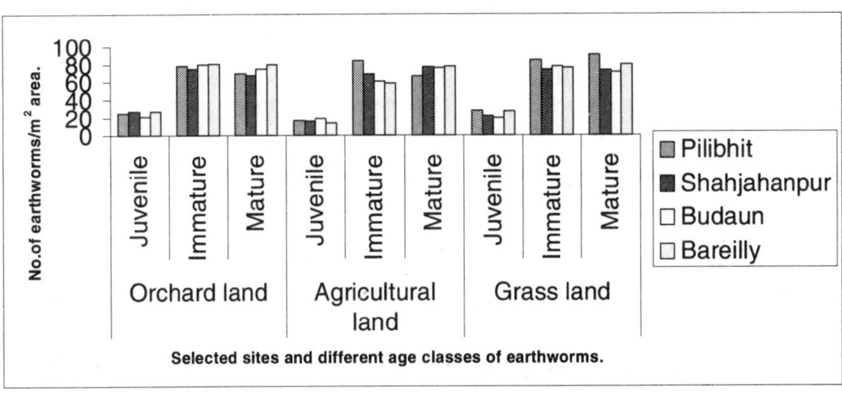

Fig.1b: Population dynamics of earthworms in three agro-ecosystems of Bareilly region.

Low range of atmospheric and soil temperatures and more % relative humidity, % moisture content of soil were some of the significant factors responsible for having the maximum population density of earthworms in grassland area (204/ m^2) of Pilibhit district with respect to other experimental lands of the region (Table.3). It was also observed that the increased temperature of the atmosphere and soil reduced (34.8, 33°C) the population density of earthworms in agricultural land area (151/ m^2) of Bareilly district.

Table: 3 - Climatic factors of three study sites of the region during rainy season

Sites	Districts of the Bareilly region	Atmospheric temp. (°C)	Relative humidity (%)	Soil temp*. (°C)	Soil moisture* (% g)
Orchard land	P	32.5	86	29.0	32.45
	S	32.0	85	29.5	30.21
	B	31.8	85	28.5	33.88
	BLY	31.6	82	28.8	35.05
Agricultu-ral land	P	34.5	80	32.5	30.68
	S	33.6	80	32.0	29.46
	B	33.5	76	32.6	26.87
	BLY	34.8 max.	75 min	33.0 max.	22.34 min.
Grass land	P	29.0 min.	98 max	28.0 min.	38.88 max.
	S	30.0	84	29.8	28.81
	B	30.8	86	29.0	30.20
	BLY	29.8	80	28.5	36.10

Index: P = Pilibhit, B = Budaun, S = Shahjahanpur, BLY = Bareilly

*Soil temperature and moisture taken at the depth of 10 cm from the surface

Physico-chemical parameters of the soils are also the controlling factors to decide the worm's population. Soil having higher percent of organic matter has more population of earthworms than that of the soil with a low percent (Table 4). Further, it may be added that worm's activity increases the water holding capacity of the soil. It is fascinating to note that the worm-castings have lower pH level than that of the soil of agricultural and grass lands of the region. Singh (1989), while working on the eco-physiology of two local earthworms, *Metaphire posthuma* and *Eutyphoeus waltoni*, has also reported that the castings pH is always lower than that of the soil where the worms live. However, it was more in the soils of orchard lands of the region.

Earthworm species diversity

Four genera with seven species of earthworms viz. *Metaphire posthuma, Megascolex mauritii, Perionyx excavatus, Eutyphoeus waltoni, E.paivai, E. gigas and E. pharpingianus,* have been identified belonging to the family Megascolecidae from the region (Table 5). All the species of earthworms were abundantly present in all the three eco-systems except *E.paivai, E. gigas* and *E. pharpingianus. M. posthuma* and *E. waltoni*

Table: 4 - Physico-chemical parameters of soil and worm castings of three agro ecosystems of Bareilly region.

Agro-ecosystems	Districts of the region	Soil pH	Castings pH	% OM (Soil)	% OM (Castings)	% WHC (Soil)	% WHC (Castings)
Orchard land	P	7.12	7.39	3.32	5.28	43.37	47.26
	S	6.78	7.02	3.51	5.54	35.32	38.64
	B	7.46	7.89	3.64	5.97	49.50	53.35
	BLY	7.33	7.93	3.72	5.68	47.21	51.32
Agricultur-al land	P	8.84	7.94	2.83	3.81	42.95	45.26
	S	7.73	7.12	2.83	4.29	41.36	45.66
	B	8.45	7.90	1.95	3.51	39.36	43.76
	BLY	8.46	8.01	1.75	3.71	37.22	39.24
Grass land	P	8.81	8.62	4.02	5.90	48.43	52.46
	S	8.09	7.93	2.85	4.98	46.06	48.84
	B	7.55	7.38	2.86	4.60	45.81	48.13
	BLY	8.12	7.94	3.83	5.78	38.20	42.67

Index: OM = Organic matter, WHC = Water holding capacity

Table: 5 - Taxonomic position of collected species of earthworms from the region

Family-Megascolecidae	
Sub-family: Megascolecinae	**Sub-family: Octochaetinae**
i) *Megascolex mauritii* (Kinb.).	i) *Eutyphoeus waltoni* (Mich).
ii) *Metaphire posthuma* (Vaill.)	ii) *E.gigas* (Steph.).
iii) *Perionyx excavatus* (Perr.).	iii) *E.paivai* (Mich.).
----------	iv) *E. pharpingianus* (Mich.).

were found in all the districts of the region. *M. mauritii* was recorded from all the districts except the orchard and agricultural lands of Shahjahanpur district. Availability of *P. excavatus* was noticed in orchard and grassland of Pilibhit and Bareilly districts. However, this species was also found from the orchard land of Shahjahanpur district. Density of all the species of *Eutyphoeus (excluding E.waltoni)* was very rare. *E.paivai* was reported from orchard land of Pilibhit district, *E.gigas* from the agricultural land of Pilibhit and *E.pharpingianus* from agricultural land of Pilibhit and Badaun and grassland from Bareilly districts (Table 6).

Population dynamics

During rainy season juveniles and immature dominated over the whole population in all the three sites (Fig. 1b). It may be due to their active

Table: 6–Showing species diversity of earthworms in three agro-ecosystems of Bareilly region

Earthworm species	Orchard land				Agricultural land				Grass land			
	P	S	B	Bly	P	S	B	Bly	P	S	B	Bly
1. *Metaphire posthuma*	*	*	*	*	*	*	*	*	*	*	*	*
2. *Megascolex mauritii*	*	—	*	*	*	—	*	*	*	*	*	*
3..*Perionyx excavatus*	*	*	—	*	—	—	—	—	*	—	—	*
4. *Eutyphoeus waltoni*	*	*	*	*	*	*	*	*	*	*	*	*
5. *Eutyphoeus paivai*	*	—	—	—	—	—	—	—	—	—	—	—
6. *Eutyphoeus gigas*	—	—	—	—	*	—	—	—	—	—	—	—
7. *Eutyphoeus pharpingianus*	—	—	—	—	*	—	*	—	—	—	—	*

* = Present, — = Absent

breeding period (Evans and Guild, 1948; Edwards and Lofty, 1972; Raw, 1962; Satchell, 1967; Dash and Patra, 1977 and Lavelle, 1978).

Factors influencing the diversity and abundance of earthworms

Physical and organic factors of soil have been known to influence the abundance and distribution of earthworms alongwith the climatic conditions of the study sites (White, 1975). Dowdy (1944), Evans and Guild (1947), Grant (1956), Dash and Patra (1977) have reported that soil moisture and temperature are important factors regulating the population of earthworms. Edwards and Lofty (1972) have reported that the pH of the soil limits earthworm distribution. Gupta and Sakal (1967) have estimated that the pH of garden soil was lower than that of castings as in case of soils of orchard lands of the region in the present studies. However, it was different in agricultural and grassland areas of the region indicating that some unknown factors also affect earthworms activities. They have further reported that the worm-castings have higher amount of organic matter than the soil, similar findings as reported in the present work. Amount of organic matter available in a particular type of soil may be a hot spot area that determines the population density and diversity of earthworms. The maximum population density and species diversity of earthworms have been reported in the soils having the organic matter in the range of 2.85-4.02% in the present studies.

Segregation and selection of available earthworm species on the basis of their habitat for vermicomposting

Perionyx excavatus was found in the areas having dense amount of organic matter in the orchard lands of Pilibhit, Shahjahanpur and Bareilly districts of the region. However, it was present in the grassland area of Bareilly and Pilibhit districts and it may be used in solid waste management. Other species of earthworms reported during the study are basically endogeic and fit for soil improvement.

Conclusion

Pilibhit district of the Bareilly region is one of the hot-spot areas for earthworms representing their maximum diversity in all three study sites; while Shahjahanpur, the minimum. Decreasing sequential diversity pattern of earthworms in all the four districts of the region during the study was as follows:

Pilibhit < Bareilly <Badaun < Shahjahanpur

Two species of earthworms, *Metaphire posthuma* and *Eutyphoeus waltoni* are the commonest species reported in all three study sites from the districts of the region. Although, *Megascolex mauritii* is also one of the local species but not found in the orchard and agricultural lands of Shahjahanpur district.

References

Agrawal, D., and Agrawal, O.P.2008: A study on the biodiversity of earthworms in and around Gwalior (Madhya Pradesh). In Earthworm Ecology and Environment .(Edi, SM Singh) IBDC Publication, India. P 15-24.

Chaudhuri, P.S. and Bhattacharjee, G. 1999: Earthworm resources of Tripura. Proc. Nat. Acad. Sci. India,69 (B), II, 159-70.

Dowdy, W.W.1944: Influence of temperature on vertical migration. Ecology, 25 (4): 449-460.

Dash, M.C. and Patra, U.C. 1977: Diversity, biomass and energy budget of a tropical earthworm population from a grassland site in Orissa, India. Rev. Ecol. Biol. Sol. 16: 79-83.

Edwards, C.A. and Lofty, J.R.1972: Biology of earthworms, I ed., Chapman and Hall Ltd., London. p. 283.

Edwards, C. A. and Lofty, J. R. 1977: Biology of earthworms, II ed. Chapman and Hall Ltd., London, p. 333.

Evans, A. C. and Guild, W.J. Mc. L. 1947: Studies on the relationships between earthworms and soil fertility. Biological studies in the field. Ann. Appl. Biol. 34 (3): 307-30.

Evans, A.C.1948: Relation of worms to soil fertility. Discovery. Norwich 9 (3): 83-86.

Grant, W.C.1956: An ecological study of the peregrine earthworm, *Pheretima hupeiensis* in the Eastern United States. Ecology, 37(4): 648-658.

Gupta, M.L. and Sakal, R.1967: A comparative study of the physico-chemical composition of earthworm castings of the garden sand cultivated soils. Jour. Sci. Res. Banaras Hindu Uni.17: 213-220.

Gates, G.F. 1972: Burmese earthworms. An introduction to the systematics and biology of megadrile Oligochaetes with special reference to Southeast Asia. *Trans. Am. Phil. Soc.*, **62:** 1-326.

Ismail, S.A. 1986: Earthworm resources of Madras. Proc. Nat. Sem. Org. Waste Utiliz. Vermicomp. Part - B: Verms and Vermicomposting (ed. M.C. Dash, B.K. Senapati and P.C. Mishra), Sambalpur University, Orissa: 8-15.

Julka, J.M. and Paliwal, R. 1986: Distribution of Indian Earthworms. In: *Verms and Vermicomposting*. Proc. Nat. Sem. Org. Waste Utiliz. Vermicompost. Part-B. (Ed. M.C. Dash, B.K. Senapati, and P.C. Mishra). Five Star Press, Burla, p. 16-22.

Julka, J.M. 1988: The Fauna of India and the Adjacent Countries. Megadrile :Oligochaetidae. Zoological Survey of India, Calcutta, India, p. 400.

Julka, J.M. and Senapati, B.K. 1987: Records of Zoological Survey of India: Earthworms (Oligochaeta: Annelida) of Orissa, India. Miscellaneous Publication (Occasional Paper No.92), Published by the Director, ZSI, Calcutta.

Julka, J.M. 1993: Earthworm Resource of India and their utilization in vermiculture. In: *Earthworm Resource and Vermiculture.* Zoological Survey of India, Calcutta, p. 51-56.

Julka, J.M. 1996: Annelid diversity in the Thar Desert. In: *Faunal Diversity in the Thar Desert: Gaps in Research.* (eds. Gosh, A.K. *et. al.*). Scientific Publishers, Jodhpur, India, p. 71-76.

Kale, R.D. and Krishnamoorthy, R.V. 1978: Distribution and abundance of earthworms in Bangalore, Proc. Indian Acad. Sci. 87B: 23-25.

Lavelle, P.1978: Les Vers de terre de La Savane de Lamto (cote d' I voive) Peuplemaents, populations et functions dans L' ecosysteme. Publ. Lab. Zool. E.N. Ss.12: p.301.

Michaelsen, W. 1909: The Oligochaeta of India, Nepal, Ceylon, Burma and Andaman Islands. *Mem. Indian Mus.*, **1:** 103-253.

Raw, F. 1962: Studies of earthworm population in orchards. I. Leaf burial in apple orchards. Ann. Appl. Biol., 50: 389-404.

Satchell, J.E.1967: Lumbricidae. In: Burges, A. and F.Raw (ed.). Soil. Biol. Acad. Press. London, p.259-322.

Singh, S.M.1989: Structure, development and eco-physiology of two species of earthworms. PhD thesis submitted to Lucknow University, Lucknow.p.157.

Stephenson, J. 1923: "The Fauna of British India" including Ceylon and Burma. Oligochaeta. Taylor and Francis, London.

Templeton, R. 1844: Description of *Megascolex caeruleus*. Proc. Zool. Soc., Lond. 12: 89-91.

Walkley-Black, 1947: Oxidizable matter by chromic acid with sulphuric acid heat of dilution. Soil. Sci., 63:251.

White, W. Jr. 1975: An earthworm is born (New York: Sterling Publ. Co.), p. 80.

Metaphire posthuma

Eutyphoeus waltoni

Eutyphoeus gigas

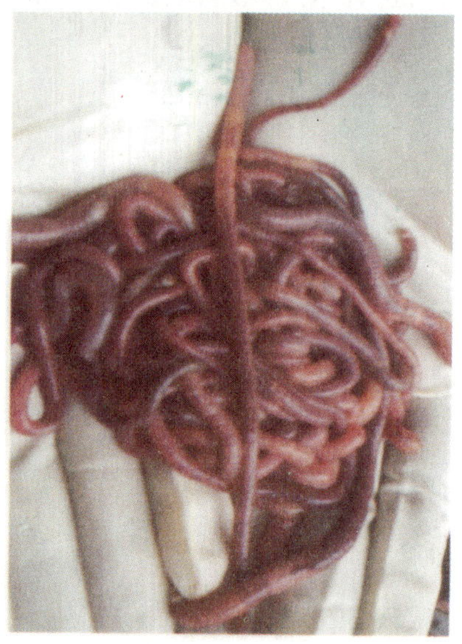

Eutyphoeus paivai

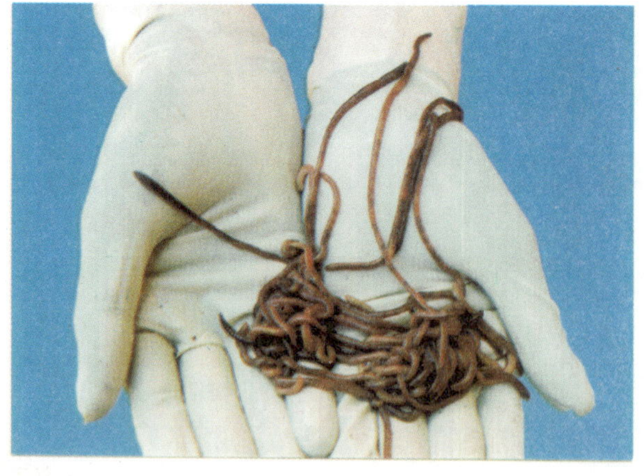

Megascolex mauritii

E. pharpingianus

Perionyx excavatus

Chapter 2

A Study on the Biodiversity of Earthworms in and around Gwalior (Madhya Pradesh)

Dheeraj Agrawal and O.P. Agrawal
Vermibiotechnology Center, School of Studies in Zoology, Jiwaji University, Gwalior-474011

Earthworms are well known invertebrate unique creatures of animal kingdom, distributed throughout the world and play an important eco-functional role in soil ecosystem by influencing physical, chemical and biological properties and productivity of the soil. Looking into their prospective role in waste management, organic farming, sustainable agriculture, environmental conservation and human health (Kale, 1998, Agrawal, 2005; Ranganathan, 2006), the biodiversity of earthworms has become one of the important lacunas to work out at different regions.

Carolus Linnaeus (1758), in his book *"Systema Naturae"* described two species of Annelids. One of them was an oligochaete earthworm, *Lumbricus terrestris.* The first earthworm species, *Megascolex caerulus* was discovered by Templeton (1844) from Srilanka - a region of Indian subcontinent. However, *Lumbricus terrestris* was the first earthworm from the Indian mainland (Perrier, 1872). Several attempts have been made to collect and identify earthworms from Indian subcontinent. Michaelson (1909) published the scattered informations on earthworm taxonomy in the form of a systematic monograph. This study had given a considerable impetus to subsequent taxonomic studies. Stephenson (1923) incorporated a consolidated taxonomic account of earthworms in *The Fauna of British India, including India, Ceylon and Burma.* A second volume followed it on Oligochaeta by Stephenson (1930). During late 19th and early 20th centuries, lot of work has been done on the morphology, histology and taxonomy of earthworms including in India. New somatic characters were included in taxonomic identification and much changes occurred in the nomenclature and the status of taxa, genera and species and hence Stephenson's documents have become outdated and obsolete. Much of the credit in this regards goes to Gates and Julka (Gates, 1972; Julka, 1988). The scientists of Zoological Survey of India have done a commendable job of conducting countrywide survey and identification of earthworms. Julka has published a number of books and articles on earthworms especially on taxonomy (Julka, 1976; 1981, 1988, 1993, 1996, 2001; Julka and Halder, 1977; Julka and Paliwal, 1986, 1994; Julka and Senapati, 1987; Julka *et al.*, 1997).

According to Reynolds (1994) there are 3,627 terrestrial species of earthworms known in the world. He discussed global distribution, barriers to migration, habitat requirement and functions of earthworms in the soil. A number of earthworms have been reported from different parts of India. Out of approximately 3600 global species of earthworms, 509 species belonging to 67 genera and 10 families have been recorded from Indian subcontinent, indicating a high degree of diversity in this region as compared to other areas (Julka, 1993, 2001). Both endemic and peregrine species are found in India.

Varma and Chauhan (1979) and Chauhan (1980) have described seven species of earthworms collected from Gwalior and studied their pH relationship and changes in seasonal activities. Julka (1981, 1988) has mentioned about five species of earthworms, collected form Madhya Pradesh, two of them from Gwalior. Asthana (1968) conducted research work on a local species of earthworm, *Perionyx cressiseptatus from Gwalior.* The present study on the biodiversity of earthworms of Gwalior has been carried out as no other work on the taxonomy and biology of earthworms has been carried out.

Collection and Dissection of Earthworms

Earthworms were collected from various habitats by hand sorting method using the techniques of Edwards and Lofty (1977) and Julka (1988). Collected worms were washed in fresh water and immobilized by the addition of a few drops of 10 % ethanol into the water. The earthworms were properly stretched and fixed in 10 % formalin for 24 hours for subsequent identification. Internal morphological characteristics of taxonomic importance were studied by dissecting preserved worms longitudinally along the mid dorsal line.

Experimental findings

Twenty-two varieties of earthworms were collected from different localities and habitats of Gwalior. Out of them 19 were collected from the soil of different habitats and 3 from vermicomposting units. The collected specimens were identified using suitable keys on the basis of external and internal morphological features and observations were recorded. Total 18 species belonging to 10 genera and 5 families have been identified. Some individuals belonging to Megascolecidae and Octochaetidae could not be identified up to the species level.

Family Megascolecidae

Eight species were identified from the family Megascolecidae including three species of genus *Metaphire* (formerly known as *Pheretima*), *M. posthuma, M. houlleti* and *M. birmanica,* two species of *Amynthas (A. alexandri* and *A. morrisi*), two species of *Perionyx (P. excavatus* and *P. cressiseptatus* and *Lampito mauritii* (Table-1). *Perionyx excavatus* was collected from the vermicomposting units. This species along with *Eisenia fetida* (family Lumbricidae) and *Eudrilus eugeniae* (family Eudrilidae) were brought from Bangalore.

Family Octochaetidae

From the collected varieties, seven species were identified to belong to the family Octochaetidae including five species of *Octochaetona,* O. *beatrix, O. paliensis, O. pattoni, O. rosea* and *O. thurstoni, Eutyphoeus waltoni* and *Pellogaster bengalensis* (Table-2).

Family - Ocnerodrilidae

In the present study a single species *Ocnerodrilus occidentalis* represented the family Ocnerodrilidae (Table-3).

Family - Eudrilidae

The specimens of giant African (Nigerian) worm, *Eudrilus eugeniae* were procured from vermicomposting units and were identified as the representative of this family (Table-3).

Family - Lumbricidae

The red wiggler worm, *Eisenia fetida,* the most popular variety of vermicomposting worm in India was also procured from vermicomposting units for identification (Table-3).

Table 1. Morphometric characteristics of different species of family Megascolecidae

Parameters	*Metaphire posthuma*	*Metaphire birmanica*	*Metaphire hauletti*	*Amynthas alexandri*	*Amynthas morrisi*	*Perionyx excavatus*	*Perionyx cressiseptatus*	*Lampito mauritii*
Length (mm)	115-140	140-145	60-70	100-290	100-113	90-103	60-80	75-150
Diameter (mm)	5-8 mm	5-6	2-3	4-20	3-4	2-4	2-4	5-8
No. of segments	90-140	110-125	104-117	100-130	90-95	110-145	132-151	90-140
Biomass (weight) mg	825-2642	1065-1072	639-725	1170-6612	798-854	160-300	200-300	825-2642
Prostomium	Epilobous	Rudimentary Epilobous	Epilobous	Rudimentary	Zygolobous	Closed epilobic	Closed epilobic	Epilobic
Clitellum	14-16 (3)	14-16 (3)	14-17 (4)	14-16 (3)	13-15 (3)	14-17 (4)	13-17 (5)	14-18 (5)
Male pore (s)	18 (paired)	18 (paired)	18/19 (paired)	18 (paired)	17 (paired)	18 (paired)	18 (paired)	18 (paired)
Female pore	14 (single)	14 (single)	14 (single)	14 (single)	13 (single)	14 (single)	14 (single)	14 (single)
Spermathecal pores	4 pairs (5/6-8/9)	3 pairs (6-8)	3 pairs (6-8)	4 pairs (6/7-9/10)	2 pairs (5/6, 6/7)	3 pairs (6/7-8/9)	3 pairs (6/7-8/9)	3 pairs (6/7-8/9)
Gizzard	8	8-10	8-10	9-10	8-9	5-6	6	7
Testes	2 pairs (10, 11)	2 pairs (10, 11)	2 pairs (10, 11)	2 pairs (10, 11)	1 pair (10)	2 pairs (10. 11)	2 pairs (10. 11)	2 pairs (10, 11)
Test. funnels	2 pairs (10, 11)	2 pairs (10, 11)	2 pairs (10, 11)	2 pairs (10, 11)	1 pair (10)	2 pairs (10. 11)	2 pairs (10. 11)	2 pairs (10, 11)
Testis sac	Unpaired (10, 11)	Unpaired (10, 11)	Unpaired (10, 11)	Unpaired (10, 11)	Unpaired (10)	Unpaired (10, 11)	Unpaired (10, 11)	Unpaired (10, 11)
Seminal vesicle	2 pairs (11, 12)	2 pairs (11, 12)	2 pairs (11, 12)	2 pairs (11, 12)	1 pair (11)	2 pairs (11, 12)	2 pairs (11, 12)	2 pairs (9, 12)
Prostate gland	17-20	17-19	17-19	18-19	17-22	18-19	17-19	18
Spermathecae	4 pairs (6-9)	3 pairs (6-8)	3 pairs (6-8)	4 pairs (7-10)	2 pairs (6-7)	3 pairs (7-9)	3 pairs (7-9)	3 pairs (7-9)
Ovary	13	13	13	13	13	13	13	13

Table 2. Morphometric characteristics of different species of family Octochaetidae

Parameters	*Octochaetona beatrix*	*O. paliensis*	*O. pattoni*	*O. rosea*	*O. thurstoni*	*Pellogaster bengalensis*	*Eutyphoeus waltoni*
Length (mm)	40-134	35-92	60-110	100-290	130-240	90-103	53-230
Diameter (mm)	2-5	2-4	2-4	3-5	5-6	2-4	4-8
No. of segments	133-197	110-190	140-160	200-280	190-235	110-145	115-201
Biomass (weight) mg	1400-1800	255-631	430-510	592-968	925-980	225-315	799-870
Prostomium	Epilobous	Epilobous	Epilobous	Epilobous	Epilobous	Tanylobous	Pro or tanylobous
Clitellum	13-17 (5)	13-17 (5)	13-16 (4)	13-17 (5)	13-17 (5)	14-17 (4)	13-17 (5)
Male pore (s)	18 (paired)	18 (paired)	18/19 (paired)	18 (paired)	18 (paired)	18 (paired)	18 (paired)
Female pore (s)	14 (single)	14 (paired)	14 (paired)	14 (paired)	14 (paired)	14 (paired)	14 (single)
Spermathecal pores	2 pairs (8-9)	2 pairs (8-9)	2 pairs (7, 8)	2 pairs (8-9)	2 pairs (8,9)	2 pairs (8-9)	1 pair (7/8)
Gizzard	4/5-8/9	4/5-7/8	4/5-9/10	9-10	4/5-7/8	5-6	7
Testes	1 pair (11)	2 pairs (10, 11)	2 pairs (10, 11)	2 pairs (10, 11)	2 pairs (10, 11)	2 pairs (10, 11)	1 pair (11)
Test. Funnels	1 pair (11)	2 pairs (10, 11)	2 pairs (10, 11)	2 pairs (10, 11)	2 pairs (10, 11)	2 pairs (10, 11)	1 pair (11)
Testis sac	Unpaired (11)	Unpaired (10, 11)	Unpaired (10, 11)	Unpaired (10, 11)	Unpaired (10, 11)	Unpaired (10, 11)	Unpaired (11)
Seminal vesicle	12	2 pairs (11, 12)	2 pairs (11, 12)	2 pairs (9, 12)	2 pairs (11, 12)	2 pairs (11, 12)	1 pair (11-14)
Prostate gland	17-19	17-19	17-19	17-19	17-19	17-19	17-19
Spermathecae	2 pairs (8-9)	2 pairs (8, 9)	2 pairs (7, 8)	2 pairs (8,9)	2 pairs (8-9)	17-19	1 pair (7/8)
Ovary	14	14	14	14	14	14	13

Table 3. Morphometric characteristics of different species of family Ocnerodrilidae, Lumbricidae and Eudrilidae

Parameters	*Ocnerodrilus occindentalis*	*Eisenia fetida*	*Eudrilus eugeniae*
Length (mm)	60-80	55-90	150-250
Diameter (mm)	2-4	4-8	5-8
No. of segments	132-151	60-95	205-230
Biomass (weight) mg	200-300	680-750	825-2642
Prostomium	Closed epilobic	Epilobous	Zygolobous
Clitellum	13-17 (5)	26-31 (6)	14-17 (4)
Male pore (s)	17 (paired)	15 (paired)	17/18 (paired)
Female pore	14 (single)	14 (single)	14 (single)
Spermathecal pores	?	2 pairs (9/10, 10/11)	3 pairs (7-9)
Gizzard	Absent	7	7
Testes	1 pair (10)	2 pairs (10, 11)	2 pairs (10, 11)
Testes funnels	1 pair (10)	2 pairs (10, 11)	2 pairs (10, 11)
Testis sac	Unpaired (10)	Unpaired (10, 11)	Unpaired (10, 11)
Seminal vesicle	1 pair (11)	9-13	2 pairs (10, 11)
Prostate gland	17-19	Absent	17
Spermathecae	?	2 pairs (9-10)	3 pairs (7-9)
Ovary	13	13	13

Suitability of local earthworms for vermicomposting

An assessment has also been made using criteria of habit, habitat, natural population density and morphological attributes, if any of the local earthworms can be used for vermicomposting. Only *Perionyx cressiseptatus* was found in clusters in moist topsoil having rich inputs of organic waste matter and it could also be collected during pre and post rainy season. Thus, it seems to be suitable for the purpose and attempts are being made to maintain its culture like other vermicomposting worms.

Status of Earthworm Diversity

Occurrence of about twenty species of earthworms from Gwalior indicates a rich biodiversity. Asthana (1968) collected *Perionyx cressiseptatus* from Gwalior to study on morphology and physiology of different organ systems. Varma and Chauhan (1979) and Chauhan (1980) have reported seven species of earthworms from Gwalior and studied their pH relationship and changes in their seasonal activities. Out of these seven species (*Pheretima = Metaphire posthuma, P. houletti, P. anomela, Perionyx millardi, P. cressiseptatus, Diplocardia singularis* and *Eutyphoeus waltoni*), only four were found in the present study (*Metaphire posthuma, M. houletti, Perionyx cressiseptatus* and *Eutyphoeus waltoni*) and the other three (*Metaphire anomela, Perionyx millardi* and *Diplocardia singularis*) were not present. This indicates that earthworm biodiversity is changing with time and space. Julka (1981, 1988) has mentioned two species of earthworms, *Octochaetona paliensis* and *Octochaetona beatrix* collected form Gwalior. However, out of other three species reported by Julka from other areas of Madhya Pradesh, only, *Pellogaster bengalensis* was found in Gwalior. Such an extensive survey of earthworms has been made for the first time in Gwalior and Madhya Pradesh. Several species are reported for the first time from this area and the state. Among the earthworm fauna of Gwalior, *Perionyx cressiseptatus, Amynthas alexandri, Eutyphoeus waltoni, Lampito mauritii, Octochaetona rosea* and *Pellogaster bengalensis* were very abundant and common.

Approximately 30 species of earthworms under 16 genera and 6 families have been reported from Orissa by Julka and Senapati (1987). Dhapola (2001) collected 10 species of earthworms from Tarai region of India (Nainital). The earthworm fauna of Tripura has been investigated by Chaudhuri and Bhattacharjee (1999, 2005) and Halder (2000). Chaudhuri and Bhattacharjee (2005) have recorded 21 species belonging to 5 families and 10 genera from pasture, compost pits and sewage sludge of Tripura. Out of these 21 species, 14 were noticed to be distributed in both Tripura and neighbouring Bangladesh (Reynolds, 1994b; Reynolds *et al.*, 1995). Ismail *et al.* (1990) have studied the earthworm fauna from selected habitat of Chennai. Twelve species of earthworms have been collected from different habitats of Bangalore (Kale and Krishnamoorthy, 1978). As compared to the earthworm fauna of Rajasthan, with a total 15 species (Julka, 1996, 2001; Tripathi and Bhardwaj, 2005), of earthworms of Madhya Pradesh appears to be richer with a total of approximately 23 species (Varma and Chauhan, 1979; Julka, 1981, 1988 and the present study). This number is likely to increase if more areas and habitats are explored.

Use of exotic varieties of epigeic earthworms for vermicomposting and waste management has often been criticized as they are likely to deceive and eradicate the local fauna or they might be the source of unknown and unidentified risks, causing plant, animal and human diseases. However, the possibility of such risk factors has been denied after critical assessment because the exotic epigeic worms have a different habitat niche than that of the local burrowing worms and during a period of two decades they have already been acclimatized in Indian conditions (Kale, 1998).

Acknowledgments

Authors are thankful to Jiwaji University, Gwalior, and Head of the School to provide funds for the present study and Prof. Kale, Bangalore, for epigeic earthworms to initiate vermiculture.

References

Agrawal, O.P. 2005: Vermitechnology for all-round sustainable development. Paper In : *National Seminar on Composting and Vermicomposting*, CSRTI, Mysore, India, p. 39-47.

Asthana, G.P. 1968: The study of morphology and physiology of the digestive, circulatory, excretory and reproductive system of *Perionyx*. *Ph. D. Thesis*, Jiwaji University, Gwalior.

Chauhan, T.P.S. 1980: Seasonal changes in the activities of some tropical earthworms. *Comp. Physiol. Ecol.*, **5:** 288-298.

Chaudhuri, P.S. and Bhattacharjee, G. 1999: Earthworm resource of Tripura. *Proc. Nat. Acad. Sci. India*, **69:** 159-170.

Chaudhuri, P.S. and Bhattacharjee, G. 2005: Earthworm of Tripura (India). *Eco. Env. & Cons.*, **11:** 295-301.

Dhapola, R. 2001: The role of earthworms on C and N cycling processes in a grassland and managed soil of Tarai region. *Ph. D. Thesis*, Kumaun University, Nainital.

Gates, G.F. 1972: Burmese earthworms. An introduction to the systematics and biology of megadrile Oligochaetes with special reference to southeast Asia. *Trans. Am. Phil. Soc.*, **62:** 1-326.

Darwin, C. 1881: *The Formation of Vegetable Mould Through the Action of Worms, with Observation of Their Habits*. Murray, London.

Edwards, C.A. and Lofty, J.R. 1977: *Biology of Earthworms*. Chapman and Hall, London, p. 333.

Halder, K.R. 2000: Oligochaeta: Earthworm. In: *State Fauna Series 7: Fauna of Tripura, Part 4. Zoological Survey of India, Calcutta, p. 239-256.*

Ismail, S.A., Ramakrishna, C. and Anzar, M.M. 1990: Density and diversity in relation to the distribution of earthworms in Madras. *Proc. Indian Acad. Sci. (Anim. Sci.)*, **99:** 73-78.

Julka, J.M. 1976: Studies on the earthworms collected during the Daphabum expedition in Arunachal Pradesh, India *Rec. Zool. Surv. India,* **69:** 229-239.

Julka, J.M. 1981: Taxonomic studies on the earthworms collected during the Subansiri expedition in Arunachal Pradesh, India *Rec. Zool. Surv. India,* Misc. Publ., Occ. Paper, **26:** 1 – 53.

Julka, J.M. 1988: *The Fauna of India and the Adjacent Countries. Megadrile :Oligochaetidae.* Zoological Survey of India, Calcutta, India, p. 400.

Julka, J.M. 1993: Earthworm Resource of India and their utilization in vermiculture. In: *Earthworm Resource and Vermiculture.* Zoological Survey of India, Calcutta, p. 51-56.

Julka, J.M. 1996: Annelid diversity in the Thar desert. In: *Faunal Diversity in the Thar Desert: Gaps in Research.* (Eds. Gosh, A.K. *et. al.*). Scientific Publishers, Jodhpur, India, p. 71-76.

Julka, J.M., 2001: Distributions of earthworms in different agro climatic regions of India. *Workshop on Tropical Soil Biology and fertility programme.* School of Environmental Sciences, J.N.U., New Delhi, p. 1-12.

Julka, J.M. and Halder, K.R. 1977: New records of earthworms (Oligochaeta Lumbricidae) from Sikkim. *Newsl.Zool. Surv. India,* **3:** 296-297.

Julka, J.M. and Paliwal, R. 1986: Distribution of Indian Earthworms. In: *Verms and Vermicomposting.* Proceedings of the National Seminar on Organic Waste Utilization and Vermicomposting part-B. (Eds. Dash, M.C., Senapati, B.K. and Mishra, P.C.), Five Star Press, Burla, p. 16-22.

Julka, J.M. and Paliwal, R. 1994: On a new species of *Plutellus* Perrier (Anacthodrilidae: Oligochaeta) from northwest Himalayas, India. *Res. Bull. Punjab Univ.,* **44:** 217-220.

Julka, J.M., Giri, S., Panigrahi, P.K. and Senapati, B.K. 1997: *Parryodrilus lavellei* gen.nov. and sp. Nov. (Octochaetidae: Oligochaeta) from Western Ghats, South India. *Eur. J. Soil Biol.,* **33:** 141-144.

Julka, J.M., Senapati, B.K. 1987: Earthworms (Oligochaeta: Annelida) of Orissa, India. *Rec. Zool. Surv. India,* Occ. Paper, **92:** 1 – 44.

Kale, R.D. 1998: *Earthworm: Cinderella of Organic Farming.* Prism Books Pvt. Ltd., Bangalore, p. 88.

Kale, R.D. 2006: *Vermicompost – Crown Jewel of Organic Farming.* Published by N.D. Kale, Bangalore, p. 248.

Kale, R.D. and Krishnamoorthy, R.V. 1978: Distribution of earthworms in relation to soil conditions in Bangalore. In: *Soil Biology and Ecology in India, UAS Technical Service.* (Eds. Edwards and Veeresh, G.K.). University of Agricultural Sciences, Bangalore, p. 63-69.

Linnaeus, C. 1758: *Systema Naturae.* Regnum Animale, **10:** 824.

Michaelsen, W. 1909: The Oligochaeta of India, Nepal, Ceylon, Burma and Andaman Islands. *Mem. Indian Mus.,* **1:** 103-253.

Perrier, E. 1872: Recherches pour server a 1' historie des *Lombriciens terrestris. Nouv. Archs Mus. Hist. Nat. Paris,* **8:** 5-198.

Ranganathan, L.S. 2006: *Vermibiotechnology: From Soil Health to Human Health.* Agribios (India), Jodhpur.

Reynolds, J.W. 1994 a: Earthworms of the world. *Global biodiversity,* **4:** 11-16.

Reynolds, J.W. 1994 b: The earthworms of Bangladesh (Oligochaeta: Megascolecidae, Moniligastridae and Octochaetidae). *Megadrilogica,* **5:** 33-44.

Reynolds, J.W., Julka, J.M. and Khan, M.M. 1995: Additional earthworms records from Bangladesh (Oligochaeta: Glossoscolecidae, Megascolecidae, Monilogastridae, Ocnerodrilidae and Octochaetidae). *Megadrilogica,* **6:** 51-62.

Stephenson, J. 1923: *The Fauna of British India including Ceylon and Burma.* Taylor & Francis Ltd., London, p 518.

Stephenson, J. 1930: *The Oligochaeta.* Clerendon Press, London, Oxford, p 978.

Templeton, R. 1844: Description of *Megascolex caeruleus. Proc. Zool. Soc. Lond.,* **12:** 89-91.

Tripathi, G. and Bhardwaj, P. 2005: Biodiversity of earthworm resources of arid environment. *J. Environ. Biol.,* **26:** 61-71.

Varma, B.R. and Chauhan, T.P.S. 1979: Preference for pH of some tropical earthworms.*Geobios,* 6:150–153.

Chapter **3**

Molecular Profiling - A Tool for Assessing Population Genetics of Earthworms

R. Meenatchi, R.S.Giraddi and D.P.Biradar

Department of Agricultural Entomology, University of Agricultural Sciences, Dharwad-580 005 (Karnataka)

Earthworms are regarded as both ecological engineers and biological sentinels and extensively used in vermicomposting and soil conservation (Spurgeon, 2003). In India, 509 species in 67 genera of earthworms are reported so far. Earthworm diagnostics traditionally relies mainly on morphological characters such as colour, shape, length, number of body segments, type of setal arrangement, position of clitellum, genetic papillae, etc. Taxonomic identification is often cumbersome because of difficulties encountered frequently in handling the diagnostic characters. Molecular profiling is nothing but sequencing technology involves comparison of nucleotide strings of known genetic regions (Hearty, 2003). Some commonly used molecular techniques to reduce problems in earthworm classification and reliable helps in characterization of earthworms and to assess the population of known and unknown taxa. Earthworm molecular profiling involves collection and preservation of earthworms, DNA isolation, PCR, characterization and differentiation with molecular makers. DNA fingerprinting acts as tool for interpretation of population.

Molecular Profiling

Strainal variations in *Eudrilus eugeniae* (Kinberg), *Eisenia fetida* (Savigny), *Perionyx excavatus* (Perrier), *Lampito mauritii* (Kinberg), *Pontoscolex corethrurus* (Perrier) and *Curgiona narayani* (Michaelson) were assessed to study the molecular profiling of these epigeic and endogeic earthworms. As the first part of the study, *Eudrilus eugeniae* from

Table 1. Genetic similarity matrix of different strains of *E. eugeniae*

Strains	Raichur	Dharwad	Nagpur	Bangalore	Coimbatore	Bijapur
Raichur	1.000					
Dharwad	0.714	1.000				
Nagpur	0.615	0.909	1.000			
Bangalore	0.615	0.545	0.400	1.000		
Coimbatore	0.615	0.545	0.400	0.800	1.000	
Bijapur	0.714	1.000	0.909	0.545	0.545	1.000

different geographical locations such as Dharwad, Bangalore, Tamil Nadu, Raichur, Bijapur and Nagpur were collected and used for DBNA isolation. Genomic DNA was isolated from different parts of the worms such as prostomium, clitellum, gut and tail part by using appropriate methodology. Ten random primers were used for the study. The amplifications were carried out on an Eppendorf master gradient cycler and the products were separated by agarose electrophoresis in 0.5x TE buffer. The primers, which produced polymorphic bands, were used for scoring present (1) or absent (0) and matrix was analyzed using NTSYSPC 2.0 software.

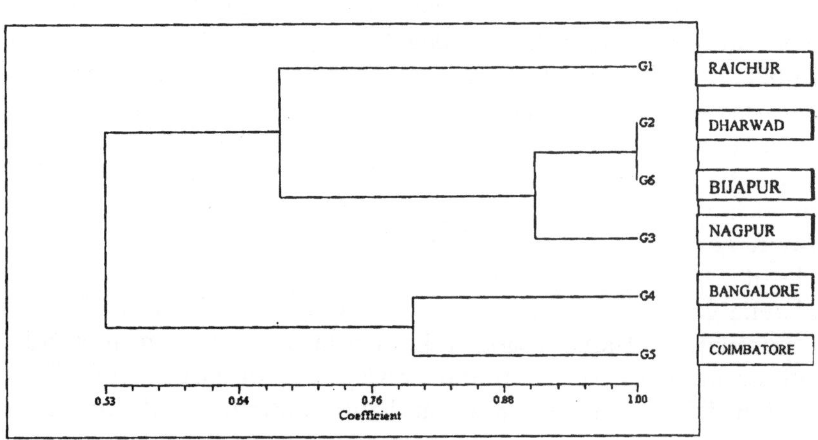

Fig. 1. Dendrogram of different strains of *E. eugeniae*

Experimental findings

The pooled molecular data and resulted dendrogram are depicted in fig.1.It formed two major clusters, with the first cluster comprising Raichur, Dharwad, Bijapur and Nagpur strains and the second cluster

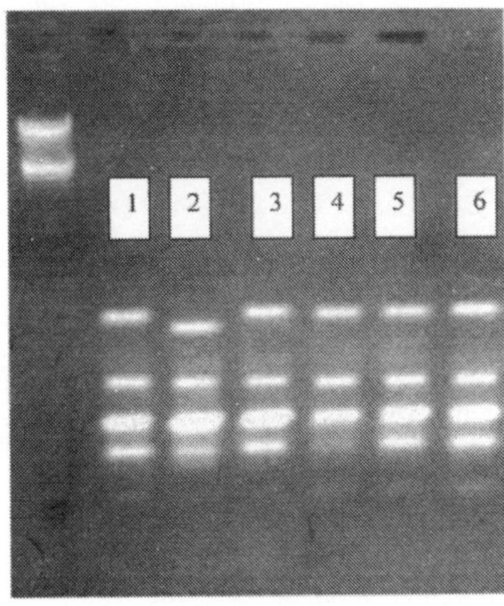

1- Raichur (Karnataka)

2- Dharwad (Karnataka)

3- Nagpur (Maharashtra)

4- Bangalore (Karnataka)

5- Coimbatore (Tamil Nadu)

6- Bijapur (Karnataka)

Fig. 3. DNA Profile of different Strains of *E. eugeniae*

with Bangalore, Coimbatore strains. The pooled matrix showed the highest genetic similarity of 100 % between Dharwad and Bijapur strains followed by 90.09 % similarity between Nagpur and Dharwad strains and Nagpur and Bijapur strains (Table.1). Highest genetic diversity of 60 % was observed between Nagpur and Bangalore strains, Nagpur and Coimbatore strains. Higher level of polymorphism with the strains of *E.eugeniae* provides valuable information for earthworm population genetics and ecological studies. The results indicated that the geographical strains from across the locations spread in three states of South India, though looked similar morphologically exhibited significant genetic variations. The molecular technique namely, RAPD-PCR proved to be an appropriate method to assess the population genetics of earthworms. The present studies are the first time report in India of molecular profiling of earthworms and throw much light on other valuable aspects of earthworm molecular systematics. Molecular diversity of earthworms viz., *L. terrestris* and *Octolasium lacteum* etc. are on record from Europe (Boore and Brown, 1995). The results indicate that various earthworm species could be validated at molecular level by screening different primers.

References

Boore, J. F.and Brown, W.M.1995: Complete sequence of the mitochondrial DNA of the annelid worm, *Lumbricus terrestris*. Genetics, 141,305-319.

Hearty, J.M.2003: Molecular systematics, Chalcidoidea and Biological control. In: Genetics, Evolution and Biological control (Eds.C.E.Ehlor,R.Siorza and T.Manteilla).CLAB International, walling Lford,U.K.,39-71.

Lspurgeon, D.J.Lweks, FJ.M.and Van Gastel Cam 2003: A summary of eleven years progress in earthworms ecotoxicolgy. Pedobiologia, 47,588-606.

PART - II

UTILITY AND SIGNIFICANCE OF EARTHWORMS

Chapter **4**

Earthworm Technology - A Promising Tool for Second Green Revolution

Priyasankar Chaudhuri
Department of Zoology, Maharaja Bir Bikram College, Agartala-799 004 (Tripura), India.

In order to make the country self sufficient in food grain production, a package programme consisting of use of chemical fertilizers, pesticides, high yielding varieties of seeds and extensive irrigation was taken in the mid-sixties of the last century. Initially this programme led to dramatic increase in food grain production, which was called Green Revolution (1967-1995). In the first phase, Green Revolution was mainly restricted to Punjab, Haryana and western part of Uttar Pradesh. Later, it was spread over to Bihar, West Bengal, Madhya Pradesh and a few other parts of the country. In fact, all those parts that enjoyed the benefit of Green Revolution were plain lands with good physical structure and irrigation facilities. Noticeably, dramatic increase in crop production was noticed up to 1985, thereafter growth rate of agricultural production had lost its momentum.

Weakness of First Green Revolution

Although Green Revolution made the country self sufficient in food grain production, it had the following drawbacks:

1. Green Revolution was restricted to some specific parts of the country. North -eastern states were totally deprived of its effects.
2. Green Revolution was centered on wheat and paddy. There was no impact on pulses - the poor man's protein.
3. In the second phase of Green Revolution (1981-1995) agricultural production was not sustainable.

4. Environment was overlooked.

Cause of decline in agricultural productivity in the second phase of Green Revolution

It is necessary to know the structural components of soil in order to find out the root-cause for decline in agricultural productivity in the second phase of Green Revolution. Soil consists of four major components: mineral materials, organic matter, air and water. Of the total soil volume, about 50% is pore space, 45% mineral matter and 5% organic matter. At optimum moisture for plant growth, the pore space is divided roughly into half; 25% water space and 25% of air. The proportions of air and water are subject to rapid and natural great fluctuations under natural conditions, depending on the weather and other factors. The four major components of a typical soil exist mainly in an intimately mixed condition. This encourages interactions within and between the groups and permits marked variation in the environment for the growth of plants (Brady, 1984).

Soil is a tremendous biological laboratory. It harvours a varied population of living organisms like bacteria (more than 1 lakh in 1 gram of soil), fungi, actinomycetes, nematodes, earthworms, micro and macro arthropods etc. Activities of soil organisms range from the largely physical disintegration of plant residues by micro arthropods and earthworms to the eventual complete decomposition of these residues by smaller organisms like bacteria, fungi etc. Accompanying these decaying processes is the release of several nutrient elements, including nitrogen, phosphorus and sulphur. Humus is one of the most useful products of microbial action.

Good physical health of soil was one of the key factors to boost agricultural production in the first phase of Green Revolution. Indiscriminate use of chemical fertilizers and pesticides for the last four decades made the tropical soil unproductive due to burning of soil organic carbon, deterioration of soil physical structure (thus affecting soil porosity and water holding capacity) and crops prone to repeated pest attacks. Agricultural revolution of the past thus turned into present days agricultural disasters. Costly chemical inputs lead not only to soil biodiversity crisis but also contamination and pollution of soil, water, air, plants and crops. The damage caused through agro-chemical pollution to environment and human health, directly and through the human food chain is irrepairable. In many cases over ninety percent of the inorganically produced vegetables, food grains, fruits, milk etc. contain poisonous agro-chemical residues harmful and unsuitable for human consumption (Paroda, 2001). The overall Indian agriculture is now in crisis due to graying of Green Revolution. So in the beginning of the twenty-first century, environmental and agricultural scientists are planning for a Second Green Revolution for sustainable agriculture (i.e. "evergreen revolution") where farming would be based largely on rejection or reduced use of chemical fertilizers, recovery of soil health and sustainable increase in food grain production through organic farming and conservation of biodiversity.

Steps to be taken in Second Green Revolution ("Evergreen Revolution")

(1) Adoption of organic farming systems

In order to make the agricultural production sustainable and keep the environment healthy, adoption of the following eco-friendly agricultural practices are recommended:

(i) **Organic manures** viz. worm-worked compost i.e. vermicompost, ordinary compost, green manures, poultry manures, blood meal, sewage sludge etc.

(ii) **Bio-fertilizers** viz. *Rhizobium, Azospirillum,Azotobacter, Phosphobacterium, VA Mycorrhiza,* Blue green algae, *Azolla* etc.

(iii) **Crop rotation with intercrop legume for atmospheric nitrogen fixation.**

(iv) **Biopesticides** viz. neem, tobacco, datura etc.

(v) **Biological Control:** The use of parasitoid, predators, pathogen and competitor population to suppress pest population.

The organic farming system has been designed for creating eco-friendly and pollution free environment, food security and safety for improving public health, ecological balance and micro-environment for growth of soil micro flora and fauna and aboveground vegetation.

Requirement of manpower is much more in organic farming system unlike machinery-based costly chemical farming practices. So scope of economic upliftment of poor class farmers is much more in Second Green Revolution.

(2) Biodiversity Conservation

Indiscriminate use of chemical fertilizers and pesticides during First Green Revolution leads to biodiversity crisis. Only 1% of pesticides act on target organ and the rest through soil bring deleterious effect on soil biodiversity. Overdoses of pesticides cause decline in the population of earthworms, mites, springtails and nitrogen fixing bacteria. All these organisms are in some way associated with improvement of physicochemical status of soil. So instead of chemical pesticides, biopesticides have been recommended for evergreen revolution.

(3) Emphasis on dry and tilla land farming

Dry and tilla lands of India were deprived of benefits of the First Green Revolution. Organic farming improves physical structure of the soil through increase in soil aggregation and porosity, thereby increasing the water holding capacity of the soil. Slow release of nutrients from organic manure leads to their proper utilization by plants and sustainable agriculture. It is expected that in the hilly and dry areas of India, crop production could be increased to a large extent following adoption of organic farming.

Earthworm–nature's best gift in organic farming

Earthworms, through their activities of feeding, burrowing and casting activities modify the physical, chemical and biological properties of soil, thus supporting above ground vegetation. Soil physical properties affected include aggregate stability and porosity, while soil biological and chemical properties modified include nutrient cycling, formation of plant-available nutrients, organic matter dynamics, pH, microbial and faunal activities, decomposition rate etc. Consequently above-ground plant production may be affected by activities of earthworms in which three main ecological groups can be recognized: **epigeic, anecic** and **endogeic.**

i) Epigeic species:

Epigeic species are litter and dung dweller, creating no burrow system in the soil, so their effects are limited primarily to the upper few centimeters of the soil-litter interface. By communiting the litter, they modify its physico- chemical status, generally reducing its C/N ratio, making it more available for microbial activity. They have small to medium body size, deep pigmentation, dorsoventrally flattened body, tolerant to disturbance, high metabolic rate, high fecundity, short life cycle and high power of regeneration. Epigeic worms form the forest floor community of temperate and tropical countries. They can be utilized for vermicomposting. Examples include *Perionyx excavatus, Eisenia fetida and Eudrilus eugeniae.*

ii) Anecic species:

Anecic and endogeic earthworms are called the "ecosystem engineers", often exerting a regulatory force in soil function. Anecics live in vertical burrow that open to the surface. They feed on decaying plant litter with some amount of soil and form surface casts. Anecics are generally large, moderate to heavily pigmented and intolerant to disturbance. Examples of anecic worms are *Drawida grandis, Eutyphoeus gammiei* etc. Anecic earthworms show good response to chemical/ phytochemical method of extraction from soil (Chaudhuri *et al.,* 1996). Temperate parts of the world are dominated by both epigeic and anecic earthworms.

iii) Endogeic species:

Polypheretima elongata, Pontoscolex corethrurus, Lampito mauritii etc. are geophagous and form both sub-surface and surface casts. They are non-pigmented or lightly pigmented, form extensive burrows (both horizontal and vertical) through the upper and lower soil horizon and establishes mutualistic relationship with the soil micro flora in their guts. Endogeic earthworms dominate tropical parts of the world. Both anecic and endogeic earthworms have long life cycle with limited power of regeneration. Endogeic earthworm species have diverse effects on soil properties. Compacting species like *P. corethrurus* egests large and compact globular casts. They increase the proportion of large aggregates in soil and bulk density. On the other hand, 'decompacting' species, *L. mauritii* feed at least partly, on large compact casts of compacting species of earthworms and egests smaller and fragile aggregates i.e. granular casts (Lavelle *et al.,* 1998). They decrease the proportion of large aggregates in soil and bulk density.

Epigeic species are concerned with humus formation in the surface soil. Both anecic and endogeic species increase overall soil porosity and enhance water infiltration. The three categories of earthworms thus contribute differently to soil fertility and structure and may be manipulated in a given area in a specific agricultural situation.

Earthworm Technology

Earthworm technology involves utilization of suitable species of earthworms to improve soil health and thereby enhancing plant growth. Earthworm technology is of two types: **in-soil technology** and **off-soil technology** (Lavelle *et al.,* 1998).

(A) In-soil earthworm technology:

This involves **direct management** as well as **indirect management** (Lavelle *et al.*, 1998).

Direct management:

In these practices, locally available endogeic and anecic earthworms, following their mass culture, are inoculated directly in the soil of plantation crop with or without proper organic input. Most successful species tested so far for in-soil earthworm technology is *Pontoscolex corethrurus*. From a culture bed of 5m x 1m x 20 cm filled with a 3:1 mixture of soil and partly composted sawdust, 200 adult *P. corethrurus* (100 g fresh weight) produced 3355 worms (including immature) within 4 months (Senapati *et al.*, 1999). In terms of effects, tree seedlings have been shown to be highly responsive to the inoculation of *P. corethrurus* in nursery bags (Lavelle *et al.*, 1998). Following application of selected species of earthworms in the tea garden at Western ghat, South India, there was significant improvement of overall biological status of soil coupled with improvement of physico-chemical characteristics of the soil, over the control plot. The impact of this technology resulted in drastic reduction (30% to 50%) in the use of inorganic fertilizers; enhancement of processed tea production between **33%** and **80%** compared with conventional inorganic fertilizer input-based management practices; and above all long-term sustainability of the soil system (Senapati *et al.*, 1999). In fact, direct management practice may positively affect plant growth only if a large earthworm biomass (at least equal to a fresh weight value of 35g m^{-2}) is inoculated from the beginning.

Indirect management:

These practices involve mulching, organic matter input, crop rotation, minimum tillage, incorporation of legumes etc. to maintain earthworm populations at a critical level of 30-40g m^{-2} (fresh biomass) resulting in a significant increase in plant production.

While in-soil earthworm technology ensures the associated soil biodiversity and long term sustainability of the soil ecosystem, it has certain limitations.

Direct management :

This would not be effective in dry lands or high rain fed zones. Besides these, due to high cost of earthworm production and inoculation, direct management practices can only be realistically applied to high value crops like tea, coffee etc. Considering cost-benefit evaluation and rapid loss of organic carbon, indirect management is preferable to direct management under tropical climatic condition.

(B) Off-soil earthworm technology: Vermicomposting

Off-soil/ex-soil earthworm technology, popularly known as vermicomposting, allows quick transformation of organic wastes into plant-nutrient rich compost called vermicompost through synergistic actions of bacteria and epigeic species of earthworms or garbage worms. Details of the process of vermicomposting under Indian conditions have been discussed by Kale (1998) and Chaudhuri (2005, 2007). The best known species with potential for waste management under Indian conditions include *Perionyx excavatus, Eudrilus eugeniae, Eisenia fetida* and *Eisenia andrei*. There is no significant difference in the quality of compost produced by *Eisenia, Eudrilus* and *Perionyx*. *Eudrilus* tops the list of vermicomposting worms when rate of vermicompost production by these species are considered. In spite of this, most commonly used earthworm species for commercial vermicomposting is *E. fetida* and its closely related *E. andrei*. This is due to the fact that *Eisenia* are generalized feeder, easily handled, a prolific breeder and have a wide range of ecological tolerance. In India *Perionyx sansibaricus* and *Perionyx pallus* have also been tested for their potential in organic waste degradation (Kale, 2004). *Polypheretima elongata, Lampito mauritii* and *Dichogaster curgensis* should not be considered as vermicomposting species because they always need soil bedding for their survival.

Management during vermicomposting

Proper management produces good quality compost within a short period. The key to maximum productivity is to maintain aerobicity in the waste combined with optimum moisture and temperature conditions. Loss of nutrients through volatilization or leaching becomes minimum during efficient and rapid vermicomposting. The quantity of water to be mixed depends upon the nature of wastes. Vegetable and fruit wastes, do not require much water because such kind of wastes give out water during the process of decomposition. When agricultural wastes (residues of sugarcane, coconut and betel-nut, straw etc.) is used, it is to be mixed with cow dung slurry to hasten up the initial decomposition. Fresh kitchen wastes, poultry and pig manures contain significant amount of inorganic salts and ammonia that may kill earthworms. Prior to earthworm inoculation, these should be removed through composting, washing or aging

(Edwards, 1998). High quality substrates like pig manures, kitchen waste, tend to decompose rapidly even in absence of earthworms. It is felt that greatest gains in containment of nutrients can be achieved by placement of bulking carbonaceous wastes (e.g. straw, paper, sawdust etc.) to increase the C to N ratio and extend the biological activity, maximizing earthworm yields, thus the proportion of nutrients to be extracted.

During the process of vermicomposting earthworms multiply. The overcrowding will reduce and hamper their growth and reproduction. So once the population reaches a peak, harvesting of worms is needed immediately. To make the harvesting process easier, vermicomposting tank should be built with placement of a longitudinally oriented porous partition (for movement of earthworms) inside. Vermicomposting is allowed in one side of the partition. After about one month of vermicomposting the empty chamber is to be loaded with partly decomposed wastes. This process allows easy movement of worms from the matured compost to new food material.

Fastest processing of organic waste is achieved by inoculation with nearly fully-grown earthworms because of their high rate of feeding activity. Culture beds should be kept free from giant flat worms, termites, rodents and centipedes.

Physical and Chemical characteristics of vermicompost

Vermicompost produced from most of the organic wastes is usually a finely divided homogeneous material with excellent structure, porosity, aeration, drainage and moisture holding capacity (Edwards, 1998). The level of macro and micro nutrients is higher than the compost derived from any other method (Kale, 1998). Most important feature of vermicompost is that, during processing of the various organic wastes by earthworms, many of the nutrients are changed into available forms that are more readily taken up by plants, such as nitrate or ammonium nitrogen, exchangeable phosphorus and soluble potassium, calcium and magnesium. So, vermicompost have a higher level of available nutrients from the wastes from which they were formed (Buchanan *et al.*, 1988). A few among many of its important properties show release of plant nutrients, improvement of soil physical properties by increasing soil aggregation, enhancement of micronutrient element nutrition of plants through chelation reaction etc. (Martin and Focht 1986), presence of plant growth regulators such as hamates and fulvates and large amount of plant growth hormones such as indole acetic acid (IAA), kinetin or gibberellins produced by bacteria within the worm castings (Arancon and Edwards, 2007).

Another important feature of vermicomposting is that during production of vermicompost, bio-available heavy metals and human pathogens are eliminated and vermicompost contains much more nitrate nitrogen and humic acid and less C : N ratio than the ordinary compost (Dominguez, 2004). Major nutrient contents in the vermicompost derived from various organic wastes are given in Table 1.

Table 1: Major plant nutrients in earthworm- processed wastes

Waste materials	Element content (% dry wt)					Reference
	N	P	K	Ca	Mg	
Cattle solids	2.20	0.40	0.90	1.20	0.25	Edwards & Burrows (1988)
Pig solids	2.60	1.70	1.40	3.40	0.55	Do
Cattle solids + straw	2.50	0.50	2.50	1.55	0.30	Do
Pig solids + straw	3.00	1.60	2.40	4.40	0.60	Do
Chick solids + wood shavings	1.80	2.70	2.10	4.80	0.70	Do
Leaves	0.80	0.34	0.82	0.96	0.53	Lofs- Homin (1985)
Sugarcane	2.67	2.11	0.40	4.08	1.89	Ramon & Romero (1993)
Banana 'stem'	2.50	0.56	3.74	2.36	1.50	Do
Kitchen Waste	1.81	1.12	0.87	2.95	0.41	Chaudhuri *et al* (2001)
Paper Waste	1.10	0.57	0.29	8.02	0.53	Picone *et al* (1987)

Vermicompost 'tea'

Watery extract obtained from vermicompost is popularly called vermicompost 'tea'. Micro organisms present in vermicompost 'tea' inactivate and suppress the growth of pathogens. Hence the extract of vermicompost can be used as foliar spray. This liquid organic manure, vermicompost 'tea', can be prepared by taking one kilogram of freshly collected vermicompost with three litres of water and is allowed to soak for two days. The mixture is thoroughly stirred before decanting to use it as spray. This preparation has the soluble nutrients, water-soluble organic substances, mucus secretions of earthworms and microorganisms. Liquid organic manure showed positive results with pot experiments (Kale, 2006). Vermicompost 'teas' also contain plant growth regulators produced by the micro-organisms (Edwards and Arancon, 2007).

Vermiwash

Vermiwash is obtained by washing of earthworms in luke warm water. It can be prepared by using the following method (Kale, 2006):

Known weight of earthworms (1/2 kg) when immersed in half a litre of luke warm water (37^0 to 40^0C) for about 30 to 40 seconds and agitated by hand, coelomic fluid oozes out through the dorsal pores. This is followed by immersion of the same earthworms in half litre water at room temperature (25^0 to 28^0C) to overcome the mild hot shock and the fluid sticking to their body mixes with the water. Water in both the trays is mixed which is vermiwash.

Utility of vermiwash

The body fluid extract of earthworm has shown the formation of inhibitory zones for some plant pathogens in petri plates at the time of application. So it can be used to spray the seedlings or for soaking seeds before sowing. Excellent results were obtained when leguminous seeds were treated with vermiwash to induce nodulation. It has also been tested as a component of nutrient media in tissue culture and has shown good response to callus growths (Kale, 2006). Stimulatory effects of root induction in cuttings has also been recorded.

Vermicompost as plant growth medium

Figure 1 summarizes physical, chemical and biological changes brought about by application of vermicompost on the soil. Main effects of vermicompost on plant growth are due to improvement of soil on physical structure and properties (aggregation, porosity, water holding capacity), presence of biologically active metabolites such as plant growth regulators, control of plant pathogens and plant parasitic nematode population, increased microbial population and to a greater availability of mineral nutrients to plants (Dominguez, 2004). Slow release of nutrients from vermicompost has a sustainable effect on agriculture.

Vermicompost has been shown to promote growth of a wide variety of cereals, vegetables, fruit plants, ornamental plants (Kale, 2006). Following application of vermicompost on summer variety wet land paddy at north Bangalore, Kale *et al.*, (1992), observed increase in

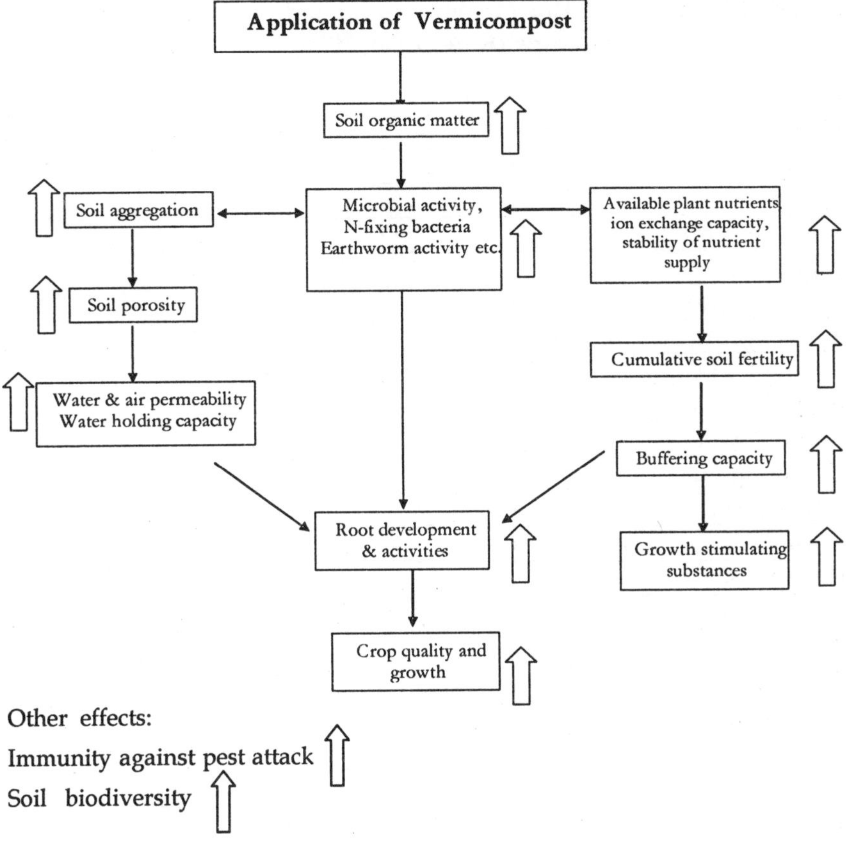

Fig-1: Physico-chemical changes in the soil brought about by the application of Vermicompost

population of beneficial microbes like N-fixing bacteria and mycorrhizae over control plots. In Tripura, a field trial conducted on upland rice (var. TRC-87-281) using 10 tons vermicompost/hectare and 5 tons of vermicompost plus NPK/hectare revealed significant increase in both grain and straw yield coupled with improvement in soil aggregation, water use efficiency and nutrient uptake in vermicompost treated plots compared to the control and NPK treated plots (Bhattacharjee et al., 2001). Application of vermicompost reduces the doses of NPK in crops. Vermicompost along with judicious use of chemical fertilizers will not only bring down the cost of cultivation but also present unique opportunities for sustainable agriculture. Opinion of farmers on using vermicompost for different crops is given in table 2.

Table 2: Opinion of the farmers about using vermicompost for different crops (Kale, 1998, 2006)

Crop variety	Opinion	Doses
(A) Cereals		2t / ac
1. Jowar	+ + +	
2. Rice	+ + + +	
3. Maize	+ + +	
(B) Pulses		2t / ac
1. Garden pea	+ + +	
2. Blackgum	+ + +	
3. Country bean	+ + +	
(C) Oil seeds		3-5t / ac
1. Sun flower	+ + +	
2. Ground nut	+ + +	
3. Soyabean	+ + +	
4. Mustard	+ + +	
(D) Spices		4t / ac
1. Cardamom	+ + +	
2. Peeper	+ + + +	
3. Curry leaf plant	+ + + +	
4. Turmeric	+ + +	
5. Clove	+ + +	
(E) Vegetables		4-6t / ac
1. Cabbage	+ + + +	
2. Raddish	+ + + +	
3. Carrot	+ + + +	
4. Potato	+ + + +	
5. Tomato	+ + + +	
6. Chillies	+ + + +	
7. Pumpkin	+ + + +	
8. Ribbed gourd	+ + + +	
9. Cucumber	+ + +	
10. Sweet potato	+ + +	
(F) Fruits		2-3 Kg/plant
1. Mango	+ + +	
2. Banana	+ + + +	
3. Water-melon	+ + + +	
4. Lemon	+ + +	
5. Grapes	+ + + +	

6. Jack fruit	+ + +	
7. Pomegranate	+ + + +	
8. Custard apple	+ + + +	
(G) Ornamental Plants		4t / ac
1. Roses	+ + + +	
2. Chrysanthemum	+ + + +	
3. Orchids, Vanilla	+ + + +	
4. Balsam	+ + +	
5. Marigold	+ + + +	
6. Lady's lace	+ + + +	
7. Tube rose	+ + + +	
(H) Cash Crop		5t / ac
1. Coffee	+ + +	
2. Tea	+ +	
3. Mulberry	+ + + +	
4. Cotton	+ + +	
5. Sugarcane	+ + + +	
6. Beetle leaf	+ + + +	
(I) Plantation Crop		5 kg / plant
1. Coconut	+ + + +	
2. Areca nut	+ + + +	
3. Teak	+ + +	

+ + + +(Excellent)   + + + (Very good)   + + (Good)   + (No difference)

Conclusion

The Green Revolution in India is the result of intensive agriculture with the extensive use of chemical fertilizers. Through intensive agricultural practices farmers could harvest three crops in a year with good irrigation facilities. The weather conditions and the soil type in the tropical countries do not favour the restoration of carbon in the soil unlike temperate countries. Continuous and indiscriminate use of chemical fertilizers increases their leaching because of the depletion in organic carbon in tropical soil. Leaching of chemicals encourages the accumulation of salts that changes the pH of the soil. Besides these, the crust formed in soil affects its porosity and water holding capacity so that soil becomes unproductive.

Repeated application of organic manure replenishes organic carbon to the impoverished soils. Presence of high level of oxidizable organic carbon helps in the slow release of nutrients from the manure and checks the leaching of nutrients. Recently, Thakur and Sharma (2005), reported that the yield, total production, income and profit from crops like maize, wheat, rajmash and peas increased significantly by two to three times under organic farming system as compared to inorganic farming system over years.

Physical, chemical and biological characteristics of vermicompost influence the growth and yield of crops. Vermicompost, being a stable fine granular organic matter, when added to clay soil loosens the soil and provides aerobicity increasing porosity. The mucus associated with the cast being hygroscopic absorbs water and prevents water logging and improves water-holding capacity. Similarly, in the sandy soils where there is a problem of water retention, the strong mucus coated aggregates of vermicompost hold water for long time (Kale, 1998). The organic carbon in vermicompost releases the nutrients slowly and steadily into the system and enables plants to absorb these nutrients. Vermicompost contains balanced nutrients and provides plant growth promoting substances that are not found in chemical fertilizer. The local soil dwelling earthworms that get established following application of vermicompost contribute to the structure and turnover of soil (Kale, 1998). Repeated application of manure to the soil of agricultural land or even wasteland thus leads to its sustained fertility. Issues of sustainability could have a cascading impact from a plot level right up to ecosystem or land scape levels, since soil fertility often is a critical limiting factor for land use development (Ramakrishnan, 2007).

Organic farming may not lead to higher production and income in the short run as its returns are of long-term nature. Organic farming system ensures in-built capacity to maintain and increase soil health and fertility leading to sustained increase in yield and production; this results in stabilization and a high jump in income and sustainable agriculture for food security in the long run (Thakur and Sharma, 2005).

Vermicomposting has gained a momentum due to present days increasing demands on organically grown produce which are free from toxic chemicals, more tasty, highly nutritious, healthy, safe and of fresh quality. Being eco-friendly, it is economically viable and a simple technology. This has helped the farming communities of Karnataka (South India) to use it in their own land to improve the fertility status of the soil and also to make additional income by selling the vermicompost produced in excess. This has resulted in improving the standard of living of these farmers (Kale, 2007). Although farmers, traders and consumers of Karnataka have followed the "organic way of life", in most of the other states of India, this is not the case. Because farmers in those states are now facing some pertinent constraints and problems viz. lack of technical know-how of scientific soil and nutrient management, problems in pest and disease control, scarcity of FYM and other organic manures. Thus there is immediate need for dissemination of the concept and knowledge of scientific agriculture to create more awareness among farmers for its success and all round benefits. In fact, whole human society will benefit immensely if the farming communities of our country switch over to extensive production and use of vermicompost (Kale 1998).

First Green Revolution was restricted to few crops of some specific plain areas with proper irrigation facilities. Application of vermicompost improves physical structure, increases the water holding capacity and nutritional status of the soil. So the vast dry and unutilized tilla lands of India need to come under organic farming. Through genetic engineering, scientists have to develop high yielding variety of seeds, which would be effective in dry land farming. Crops irrespective of types and nature of land (plain or upland) show very good/excellent response in yield following application of vermicompost. Thus, in second Green Revolution, India's gross agricultural production would seemingly be much more than that of First Green Revolution.

References

Arancon, N.Q. and Edwards, C.A. 2007: The Utilization of Vermicompost in Horticulture and Agriculture. Indo-US Workshop on Vermitechnology in Human Welfare, Coimbatore (Abstract) p.23.

Bhattacharjee, G., Chaudhuri, P.S. and Datta, M. 2001:Response of paddy (var. TRC- 87-251) crop on amendment of the field with different levels of vermicompost. Asian J. Microbiol. Biotech. & Env. Sc. 3: 191-196.

Brady, N.C. 1984: The Nature and Properties of Soils. Macmillan Publishing Co., Inc p. 13.

Buchanan, M.A., Russell, E. and Block, S.D. 1988: Chemical characterization and nitrogen mineralization potentials of vermicomposts derived from differing organic wastes. In: Earthworms in Environmental and West Management, C.A. Edwards and E.F. Neuhauser (Eds). SPB Acad. Publ.The Netherlands, p. 231-239.

Chaudhuri, P.S., Nanda, D.K. and Chaudhuri, D. 1996: Extraction of Octochaetid earthworms, *Eutyphoeus gammiei* using an aqueous extract of *Polygonum hydropiper* Linn, with a comparison of other chemical methods for estimating earthworm population. The Philippine J. Sci. **125:** 227 - 234.

Chaudhuri, P.S., Bhattacharjee, G., Pal T.K. and Dey, S.K.2001: Chemical characterization of kitchen waste vermicompost processed by *Perionyx excavatus*. Proc. Zool. Soc. Cal. 54: 81-83.

Chaudhuri, P.S. 2005: Vermiculture and vermicomposting as biotechnology for conversion of organic wastes into animal protein and organic fertilizer. Asian J. Microbiol. Biotech. Env. Sc.7: 359-370.

Chaudhuri, P.S. 2007:Vermicomposting as biotechnology for conversion of organic wastes into organic fertilizer and animal protein. In: Earthworms for Solid Waste Management, S.M. Singh (Ed.). International Book Distributing Co., Lucknow.p.75-88.

Dominguez, J. 2004: State-of-the art and New Perspectives on Vermicomposting Research. In: Earthworm Ecology, C.A. Edwards (2nd Ed). CRC press LLC, Florida.p. 401-424.

Edwards, C.A.1998: The use of earthworms in the breakdown and management of organic wastes. In: Earthworm Ecology, C.A. Edwards (Ed). CRC press LLC, Florida, p. 327-354.

Edwards, C.A. and Burrows, I. 1988: The potential of earthworm compost as plant growth media. In: Earthworms in Environmental and Waste Management, C.A. Edwards and E.F. Neuhauser (Eds). Academic Publishing, The Netherlands, p. 211-220.

Edwards, C.A. and Arancon, N.Q. 2007: The Science of Vermiculture: The use of Earthworms in Organic waste management. Indo-US Workshop on "Vermitechnology in Human Welfare", Coimbatore (Abstract) p. 12.

Kale, R.D., Mallesh, B.C., Bano, K. and Baggaraj, D.J. 1992:Influence of vermicompost application on the available macronutrients and selected microbial population in a paddy field. Soil Boil. Biochem. 24: 1317-1320.

Kale, R.D. 1998: Earthworm: Cinderella of organic farming, Prism Books Pvt. Ltd., Bangalore.

Kale, R.D. 2004: The use of earthworms: Nature's gift for utilization of organic wastes in Asia. In: Earthworm Ecology, Ed. C.A. Edwards (2nd Ed.) CRC Press LLC, Florida.

Kale, R.D. 2006:Vermicompost - crown jewel of organic farming (A memoir). N.D. Kale, Malleswaram, Bangalore.p.22.

Kale, RD. 2007:The future of earthworm research. The 1st National Symposium on Earthworm Ecology & Environment. Bareilly (Abstract V.), p. 23.

Lavelle, P., Barois I., Blanchart, E. 1998: Earthworms as a resource in tropical agroeco system. Nature and Resources **34:** 26-41.

Lofs - Homin, A. 1985:Vermiculture present knowledge of the art of earthworm farming. Institute Ecologie Miljovard, Swedish University Agronomic Sciences.

Martin, J.P. and Focht, D.D. 1986: Biological properties of soils. In: Soils for Management of Organic wastes and Wastewater. ASA, Madison, Wisconsin, p. 11-169.

Paroda, R.S. 2001 "99 per cent Pesticides remain in Environment". The Tribune, July 19.

Piccone, G., Biasiol, B., Deluca, G. and Minelli, L. 1987: Vermicomposting of different organic wastes. In: Compost: Production, Quality and Use, De Bertoldi, M.P. Ferranti, P. Hermite and F. Zucconi (Eds). Elsiver, London, p. 818-821.

Ramakrishnan, P.S. 2007:Sustainable mountain development: The Himalayan tragedy. Current Science **92** (3): 308-316.

Raman, C.J. and Romero, G.A. 1993:La Lombricultura una opeions tehnolo gica. In: Majenoy Disposicio in Final de Residuos So lidos Municipales. Congreso Regional del Sureste, SMISAAC Y AIDIS, Merida, Mexico, p.110-117.

Swaminathan, M.S. 1996: Sustainable Agriculture: Towards Evergreen Revolution, Konark Publishers Pvt. Ltd. Delhi.

Thakur, D.S. and Sharma, K.D. 2005: Organic Farming for Sustainable Agriculture and Meeting the Challenges of Food Security in 21st Century: An Economic Analysis. Ind. J. Agri. Econ. **60** (2): 205-219.

Chapter 5

Significance of Earthworms in Past, Present and Future Perspectives

O.P. Agrawal
School of Studies in Zoology, Jiwaji University, Gwalior - 474011 (Madhya Pradesh)

Earthworms are unique creatures of animal kingdom normally live in burrows or tunnels made in the soil and eat soil mixed with dead and decaying organic matter. They possess several special features, metameric segmentation, chitinous hook like structures, setae or chaetae, absence of true appendages, eyes and other sense organs, hermaphroditism, (male and female reproductive system present in the same individual) but need a partner for sexual reproduction, presence of ring like band, the clitellum which produces egg case, cocoon. There are approximately 4000 species of earthworms in the world and about 500 of them have been reported from India. The beneficial role of earthworms, particularly for agriculture has been well known since ancient times. But in modern civilization their importance declined. Now in recent past, it has again been realized that earthworms can do wonders for human society. They can help us in variety of ways: waste management, environmental conservation, organic farming, sustainable agricultural development, community health, science and technology. In broad sense, use of earthworms for human welfare can be termed as Vermitechnology. Present article reflects some light on their significance in past, present and future perspectives.

Need of earthworms and vermitechnology

Rapid development in science and technology resulted gradual progress in human civilization. In order to provide sufficient food for increasing human population, vast forest areas have been destroyed to convert them in agricultural fields and modern tools of agriculture (mechanization and chemical fertilizers and pesticides) have been implemented to increase crop production. Several anthropogenic activities such as industrialization, urbanization, excessive exploitation of natural resources, production and use of synthetic substances such as plastics, fibres, fabrics, fertilizers, pesticides, pharmaceutics, food additives, preservatives and other molecules, have been followed in non-judicious and ill-planned manner to make life more and more comfortable and to increase longevity. The first and foremost influence of modern civilization was deterioration of the environmental conditions due to pollution. The equilibrium of nature has been significantly disturbed. Several climatic, ecological and biological changes have been reported and deleterious effects such as green-house effect, global warming, ozone layer depletion, enhancement of hazardous solar radiations, increase in sea water level, nature's imbalance, soil erosion, silting, depletion and irregular distribution of rains, depletion of ground water table, increasing frequency of natural (like earth quakes, Tsunami) and man-made (Chernobel and Bhopal Gas tragedy) calamities, declining soil fertility, uncertainty of crop production, decline in the quality of produce due to toxic residues, decline in biodiversity, human health hazards etc. have been noticed.

The second important effect is several fold increase in the variety and amount of waste substances including both biodegradable and non-biodegradable. Safe and eco-friendly disposal and management of domestic, agricultural and industrial wastes are global problems today. These are the major sources of environmental pollution, foul smell, filthy and unhygienic atmosphere, spread and propagation of diseases caused by micro organisms. The usefulness of waste biomass has been ignored. These became an additional source of pollution. Natural phenomenon of waste recycling is not able to operate properly due to pollution mediated decline of populations of biodegrading organisms including micro organisms, detrivorous or scavenger animals, protozoans, snails, sow-bugs, insects, earthworms, centipedes, millipedes etc. Moreover, whatever decomposer community exists, we do not favour and let them work. Wrong practices (dumping, land filling, burning) of waste disposal and management are followed which amplify the problems. Not only much more pollution is created, the waste biomass is destroyed which could otherwise be used as resource for compost, fuel or energy. In the present scenario **"Waste can be called Misplaced Resource"**. What actually required is its correct placement, handling, processing and management.

The best solution of waste management is recycling, although it is not an easy and affordable task as we always think for bigger gains, market, profits, rapid solution and involvement of advanced technology that is not usually feasible on economic, environmental and practical grounds. Generally, we believe more on artificial means and do not wish to respect natural means. A major part (up to 80 %) of the waste consists of organic matter that can be processed or used to yield fuel, biogas, fodder and compost using biological tools.

It is now realized that the progress achieved in the field of agriculture and human health is not long lasting. The crop production has become stand-still, use of chemical fertilizers is no more resulting in enough production, farmers are committing suicide due to failure in crop yield, pest problems are not being solved by chemical pesticides due to development of resistance in them, risk to human health is increasing due to accumulation of more toxic residues in soil, air, water and food, side effects of chemical pharmaceuticals and development of drug resistance in pathogenic organisms. The quality of the atmosphere is deteriorating and posing several threats to human health. Thus three sectors viz. agriculture, environment and human health could be identified in which sincere considerations are to be paid to save and secure future of our race. Vermitechnology is one of the biotechniques to solve such problems and to contribute in the mission, **"nurture nature for our future".**

Earthworms and agriculture

The importance of earthworms has been realized since prehistoric times and their importance in agriculture and human health has been mentioned in ancient literature. Earthworms have drawn attention of philosophers and naturalists. As early as 384-322 B.C. the famous Greek philosopher, Aristotle described the importance of earthworms in agriculture and called them **"The Intestine of Earth"**. *Carolus Linnaeus* (1758) adopted zoological nomenclature for earthworms and listed two species of Annelida in the 10th edition of his famous book entitled, **"Systema Naturae"**. It was Sir Charles Darwin, who first made scientific observations on the role of earthworms in breakdown of organic debris on the soil surface and in the process of soil turnover. He studied earthworms for 40 years (1809-1882) and estimated that an acre of British farmland included approximately 50,000 worms producing about 18 tons of worm casts (excreta) per year. Earthworms are the masters of soil invertebrate community. They work as nature's unpaid labour force contributing significantly to the physical, chemical and microbial environment of the soil. Earthworms thrive best in soils that continuously remain moist and well aerated with ample organic residues. He described earthworms in his book, **"The formation of vegetable mould through the action of Worms"** (Darwin 1881) and expressed his opinion that **'Earthworms have played a most important part in the history of World than any other organism'**. No sincere attention has been paid to understand and exploit the usefulness of earthworms for more than a century.

Impressed by the extensive scientific works of Charles Darwin, Dr. Thomas J. Barrett also started his intensive study of harnessing the earthworms and using them to improve barren dry lands into fertile lush farms. He referred them as "Earthmaster Systems" and stated that selected species of earthworms can be raised on large scale to be introduced to the poor agricultural land. He also considered earthworms as humus factories of manure. Barrett (1959) mentioned in his book entitled, *"Harnessing the Earthworms"* about the potential of earthworms in shaping dark coloured fertile soil (compost) and considered them to be the greatest servant of nature. The wonderful living Zoo of earthworms beneath our feet is active day and night, silently and secretly doing the job of decomposition and humification by burying organic debris and mixing inorganic and organic matters of their excreta with soil (Satchell, 1967).

In the late 19th and early 20th centuries, extensive studies have been made on the morphology, histology and taxonomy of earthworms. Much of this work has been summarized and discussed by Edwards and Lofty (1977) in a book entitled **"Biology of Earthworms"**, by Satchell (1983) in **"Earthworm Ecology - from Darwin to Vermiculture"** and by Lee (1985) in a book **"Earthworms: Their Ecology and Relationship with Soil and Land Use"**.

Earthworms constitute one of the major soil fauna. In undisturbed soil, as much as 80% of the invertebrate fauna belongs to earthworms. Large number of earthworms in the soil is indicative of its fertile status. Earthworms play a significant eco-functional role in agriculture by affecting physical, chemical and biological properties of the soil. Due to their tunneling and locomotive activities, the mineral rich soil from deeper layers is brought up and blended in topsoil (hence known as Nature's plow). The channels created by earthworms increase soil porosity, enhance circulation of air and water and provide space for ample root growth. Further, decaying litter is also mixed up in the soil for faster microbial degradation.

Earthworms utilize and consume organic matter as food and release undigested material as faecal matter commonly known as casts. They manufacture and release a sticky fluid from their bodies that assists in locomotion and respiration and protects them from enemies' particularly pathogenic microorganisms. The earthworm casts have better percolation and dispersion coefficient than the parent soil. The mucus and nitrogenous excrements of earthworms increase the concentration of nitrogen and humic acid in the casts. The casts act as binding sites for plant nutrients and they form soil aggregates and are responsible for prevention of harmful plant pathogens, fungi, nematodes and bacteria. A higher percentage of carbon, nitrogen, magnesium and calcium have been reported in the casts than in the surrounding soil. Thus the presence of earthworms in the soil increases the level of humus content (humification) and fertility of the soil. However, the type and physicochemical characteristics of the soil, quality and quantity of organic matter and presence of toxic residues of agricultural chemicals (pollutants) available in a given habitat are equally responsible for survival and efficiency of the earthworms. It has been estimated that there may be about one million earthworms per hectare; they may contribute 20 to 100 kg of nitrogen per year; may result in 30 to 200 % increase in plant productivity and may exert approximately 30 % control of plant parasitic nematodes.

The impressive point of biology of earthworms is their understanding between friends and foes. In the habitat in which they flourish, harmful, anaerobic pathological organisms do not survive because of maintaining enough aeration, neutral pH and optimum temperature. The mucilaginous secretion of earthworms showed strong antibacterial activity against certain types of bacteria (enemies) on the one hand. On the other hand, this secretion promotes the activities of aerobic bacteria (friends) that help in decomposition of the organic matter. In fact they collaborate with aerobic bacteria for their activities. Furthermore, certain plant growth promoting factors (plant hormone like substances) are also released. Thus there is friendship trio: earthworms, bacteria and plants. Different activities of earthworms can be summarized in table 1.

Table 1 - Showing different activities of earthworms in soil ecosystem

Type of activity	Subtype of activity	Actions
Physical	Miner	porosity of soil, enhance circulation of air and water
	Mixer	turns soil up-side down and minerals from the deep layers are brought to the surface
	Crusher	grinding and mixing in the gut
Chemical	Degrader	complex organic matters are degraded through digestive enzymes
Biological	Stimulator	biocatalytic action on aerobic bacteria, enhances rate of degradation by 25 %, reduces time requirement by 25 %
	Inhibitor	Inhibits and suppresses the activities of certain microorganisms
	Collaborator	release plant growth promoting factors

But the current scenario is different. Due to chemical farming (green revolution), soil fertility is declining and it is becoming sterile and depleted from beneficial flora and fauna. The quality of food products is declining due to contamination of toxic residues of chemical fertilizers and pesticides. It has been stressed that organic farming should be followed and use of chemical agents should be restricted.

Earthworms as major tool in organic farming

In system of **organic farming** use of bio-composts, bio-fertilizers, bio-pesticides and other eco-friendly agents and techniques (crop rotation, tillage practices, bioremediation) and phase-wise reduction of chemical agents is recommended. The craze of healthy organic food is increasing gradually in high-income group of society and is likely to spread in other class of people too. The concept, in fact, is not new and it resembles in many respects with pre green revolutionary pattern of agriculture using farmyard manure, green manures, and cultural practices. It is now emphasized again that the organic waste particularly animal manure and agricultural waste should be converted into organic composts. Several techniques of such bioconversions are bacterial (anaerobic /aerobic), garbage composting, NADEP composting, biogas units and vermibiotechnological procedures. The last option is most effective in terms of time, labour and cost involved. The benefits of organic farming are:

a) Increase in crop production without damaging the environment.
b) Cost effectiveness in long run.
c) Sustainable agricultural development - increase in soil fertility and crop productivity.
d) Best utilization of waste resources (cattle dung, crop residues, food and food product) using biological tools (microorganisms, flora and fauna including earthworms).
e) Improvement of the quality of agricultural produce.
f) Maintenance of **clean and green environment.**
g) Improvement of human life with declining health hazards.

What is vermitechnology?

It is the technique of using suitable varieties of earthworms in soil improvement programmes and/or large-scale composting of organic waste into compost. It is simple organism-based, cost-effective, affordable and user-friendly biotechnology. It is concerned primarily with agriculture, abatement of environmental pollution and development of rural sector. It should be considered to be an essential component of the projects dealing with biotechnology and biodiversity parks, farming of medicinal plants, industries dealing with processing and packaging of agricultural, food and food products and management of organic waste stuffs. It can play a significant role in organic farming, sustainable agriculture and environmental conservation.

Earthworms can broadly be classified into two categories, burrowing or geophagous (endogeic and anecic) and non-burrowing or detrivorous (epigeic) worms (Bouche, 1977). Both burrowing (endogeic and anecic) and non-burrowing (epigeic) types of earthworms may contribute to play significant role in achieving the goals of organic farming and sustainable agriculture. On the basis of varieties of earthworms used, two types of vermitechnologies can be recognized.

Vermiconservation biotechnology

Conservation and culture of burrowing worms under natural, *in situ* (soil) conditions with restricted inputs of chemical pesticides and fertilizers, is designated as vermiconservation technology. It is a

'Two-in-one Package', which involves soil improvement and waste recycling. If required selected varieties of earthworms can be introduced into the soil. Several methods have been adopted to inoculate earthworms in desired fields such as release of living worms, cocoons or earthworm casts collected from somewhere else, transfer of soil blocks or turf pieces. The latest approach is introduction of encapsulated earthworm cocoons (vermipods). Each capsule contains some cocoons in a state of diapauses. They can be stored in refrigerator for 2-3 months and easily be transported and can be introduced into the fields on mass scale at low and reasonable cost at the time of crop planting. Their hatching time varies from few days up to one year.

Burrowing worms have limited regeneration capacity and are lightly pigmented and are not suitable for organic farming as they:

- show seasonal activity (better performance only in monsoon and post monsoon periods) and remain in inactive (diapause) phase during rest of the time.
- have long life cycle, their food requirements, metabolic rate, reproductive potential and population density are low.
- cannot live without soil.
- show intolerance to adverse weather conditions and are unsuitable for long-term maintenance in semi-natural or artificial conditions.
- have lower survival rates due to toxic residues of agricultural chemicals and other pollutants in soil and water.
- cannot be used for large scale waste management and composting purpose and,
- a long term strategy is to be planned to conclude and visualize significant effects of this venture.

However, vermiconservation can work well in combination with organic farming particularly with ample inputs of biocompost including vermicompost. In due course of time, the population of beneficial flora and fauna including burrowing earthworms will increase and they will influence (improve) the soil fertility.

Vermicomposting biotechnology

Culture and maintenance of epigeic worms in organic waste with little or no soil, under natural (*in situ*), semi-natural or artificial (*ex situ* or off soil) conditions is called vermicomposting technology.

Cocoons, vermipods baby worms, coin and a broken vermipod

Single vermipod

Split half vermipod

Fig. 1: Showing Vermipod technology

Surface-dwelling or non-burrowing earthworms, also known as **epigeic, detrivorous, manure or compost worms** are not common in India. They are very active, have high regeneration capacity and are darkly pigmented. They live in topsoil rich in decaying organic waste, manures and compost heaps. They consume large amounts of waste (voracious eaters), their metabolic rate is high and their turnover rate is very fast (4-5 times of waste biomass daily as compared to their own weight). Their life cycle is short, breeding rates are high and in suitable medium, under optimum conditions of moisture (~ 40%), pH (around neutral), shade and aeration, they remain active throughout the year without undergoing the phenomenon of diapause and they can build up high population density within a short period of time. They can live without soil or some soil can be mixed up in their culture medium. Though, the earthworms are cold blooded (poikilotherms) animals, the conditions in vermicomposting units usually remain stable (homeothermic) for survival of the epigeic worms and the units remain functional all round the year. Decaying cattle dung is the best medium for their culture, but a wide range of waste can be recycled. The methods of vermicomposting are very simple and they require lower inputs of money and labour. Wide range of methods can be adopted as per choice and requirements like pits, tanks, surface beds, windrows, compost bins, containers, baskets, hanging units using bags, baskets (Agrawal and Agrawal, 2006 a) and discarded car tyres (Fig. 2). They can be maintained in small, medium or large-scale units.

Fig. 3: Different vermicomposting models at Vermibiotechnology Centre at University

These are very suitable and highly efficient safe, pollution-free waste recycling into high quality vermicompost. Culture of epigeic worms, in natural decomposing (bacterial) system of biomass, enhances decomposition by 25% and reduces time by 25%. As compared to dung compost, vermicompost is 5 to 7 times richer in nutrients available to the plants. In fact the worms work in collaboration with aerobic bacteria. The worms engulf degrading waste along with bacteria. Their alimentary canal serves as a bioreactor for the growth and multiplication of bacteria. In the gut and vermicast, the population of these bacteria is approximately 100 fold more than in the medium without earthworms. These bacteria are the beneficial ones and they impart series of favourable effects in agriculture. Vermicomposting practice prevents propagation of harmful anaerobic bacteria and other organisms pathogenic to plants, animals and human beings.

Vermicompost is useful as fertilizer in agriculture, horticulture, nurseries, gardening, bonsai plantations, fishponds and mushroom cultures etc. Effective liquid fertilizers, vermiwash and compost tea can also be prepared in vermicomposting units. They are used as foliar spray. Besides promoting plant growth and productivity, they also have pesticidal properties on plant pathogenic micro organisms. The practice of vermicomposting strengthens the concept of organic farming and use of organic (including vermin-) compost and biopesticides helps in production of organic foods, free from toxic constituents. Increasing number of earthworms from vermicomposting units can be used for dissection purposes in scientific laboratories, as experimental model in research institutions, as food supplement for poultry, aquarium and pond fishes and animals or for preparation of worm meal, vermiprotein and pharmaceutical ingredients for treatment and prevention of human and veterinary diseases. Vermicomposting is a low cost venture and the units soon become self-sustainable and provide extra source of income to farmers, animal breeders and unemployed personnel and helps in uplift of socio-economic status of poor people. It is a good time-pass hobby for aged people and housewives. It provides nice project themes for students for scientific programmes, exhibitions and during vacations.

Vermicomposting could be successful only when suitable epigeic earthworms are employed. Such worms should be procured from a reliable source or other vermicomposting units. The body of these worms is uniformly red or brown coloured. The tip of their tail is usually light (yellow or pinkish) and is dorso-ventrally flattened. These worms can survive well in decaying organic waste without any soil and cattle dung is favourite medium for their rearing. The waste medium is usually required a varying period of pre-composting depending on the nature and type of waste so that bacterial degradation begins and initial heat generation subsides. Certain waste materials are to be mixed with cattle dung and or with other waste additives such as paper, saw dust, husk, wood scrap etc. The vermin-units are to be located in shade and a top covering is also required, as the worms do not like light. If improper (local, burrowing) variety of earthworms are introduced in the vermi-units, they will not thrive well and will soon disappear.

In most of the vermicomposting units red wriggler worm, *Eisenia fetida* -an exotic variety imported from Germany and Netherlands is being employed in India. The other varieties used for the purpose are giant African (Nigerian) night crawler, *Eudrilus eugeniae* and Oriental (Indian) worm, *Perionyx excavatus.*

Earthworms and environment

Rapid developments in science and technology have resulted in creating remarkable landmarks in all areas of human endeavours, agriculture, health, economics, life style, thinking, moral values, transport and communication but alarming deterioration in the environmental conditions could not be ignored. The risk factor of pollution is so high that it is recommended to generate public awareness and to include people's participation in environmental quality improvement programmes. It is stressed that all efforts and corrective measures are to be promoted for environmental conservation so as to reduce pollution, restrict emission of green house gases, check global warming, reduce the rate of ozone layer depletion, promote the practice of organic movement. It is not easy to implement or rather force upon the environment improvement programmes. Unfortunately the visible impact of such effects is too meager and slow to be visualized. The environment is thus a bigger issue concerned with vermi-technology than the agriculture.

The concept of organic farming, sustainable development and vermi-biotechnology are part of environmental conservation. Earthworms - environment friendly creatures are the major component of waste biomass degrading community. They cooperate with aerobic micro organisms for waste decomposition and compost generation on the one hand and establish symbiosis with plants by providing the nutrients and growth stimulating factors, a perfect **Environment Friendly Model** on the other (Agrawal, 2005 a, b; Agrawal and Agrawal, 2006 a). The practice of vermitechnology is favouring the activities of earthworms in the welfare of mankind. Some of the beneficial environmental concerns of the technology may be summarized below as the earthworms:

i) are the excellent indicator of pollution and the quality of environment (Eijsackers, 1998).
ii) help in pollution control by reducing (a) mishandling and burning of organic waste and (b) manufacture and non-judicious use of agricultural chemicals.
iii) also act as detoxifying agents as they are able to process and detoxify pollutants like pesticide residues, heavy metals and other toxins (Talasilkar and Powar, 1998; Chaudhuri, 2005; Maity *et al.*, 2007).
iv) strengthen integrity of the nature.
v) contribute in sustainable environmental development.
vi) help in improvement of the quality of air, water and food.
vii) help in maintenance of **'clean and green environment'**.
viii) help in improvement of human life.

Earthworms and human health

Vermicomposting has a potential not only in the field of agriculture and environment, but also in the field of human health at community as well as individual level because of the following promises:

- Decline in manufacture and use of agricultural chemicals (hence lesser pollution).
- Safe and eco-friendly handling of hazardous organic waste, which are the source of pollution and pathogenicity.
- Declining populations of harmful anaerobic bacteria and pathogenic organisms.
- Production of more healthy organic food free from toxic residues of chemicals.
- Possibility of procurement of vermiprotein, worm meal and bioactive molecules (vermiceuticals) from earthworms and vermiculture units.

Earthworms as Animal and Human Food

Earthworms are important part of food chain, being eaten up by many birds, robins, wood cock, chickens, crows, frogs, tadpoles, turtles, lizards, snakes, moles, shrews, rats, centipedes, other predatory animals and even by human beings, especially by tribals in various parts of the world. According to Lawrence and Miller (1945) earthworms contain sufficient protein and hence are widely eaten up by animals. Maoris in New Zealand, Japanese, natives in New Guinea, Africa and South India consume earthworms because of their high protein content and curing potential (Julka, 1988; Ranganathan, 2006). The earthworms are able to convert 20-40 % of assimilated energy into proteins. These figures are considerably higher as compared to mammalian herbivores (1.5-3 %). They are very promising source of high quality animal protein and essential amino acids (50-70%) and other useful constituents including fat 6-10%, carbohydrates 5-21%, minerals 2-3% and vitamins. The protein and amino acid composition of earthworms is far superior to snail and fish meat. It fulfills FAO and WHO standards, particularly in terms of lysine, methionine, cysteine and tyrosine all of which are important components of human and animal feed (Singh and Rai, 1998). Not only the proteins, earthworms can also fulfill the vitamin requirement of animal feeds as they have a range of vitamins like niacin, riboflavin (B_2), pantothenic acid (B-complex), thiamine (B_1), pyridoxine (B6), vitamin B_{12}, folic acid and biotin (B-complex) (Edwards, 1985). Earthworms are quite often used as food in adventure and action sports and competitions for earthworm eating are organized and telecasted by TV channels like AXN. In Canada and US, several contests for recipes using earthworms are held (Ghatnekar *et al.*, 1998). Earthworm recipes are served in restaurants in some countries.

Vermicomposting units have potential to produce large quantities of vermicompost, vermi-wash and earthworm biomass. These live earthworms can be used as fish bait and as food for fish and poultry or can be converted into worm meal and vermiprotein for use as concentrated feed for poultry, cattle and pets. Edwards and Niederer (1988) and Kale (1998) described the methods of preparation of worm meal. It has tremendous market in aquaculture systems of shrimps, carps, catfishes and tilapias. Worm protein has very good potential for human consumption too.

The suitability of cultured *Perionyx excavatus* in poultry and fish feed as a better animal protein source for *Tilapia* than other fishmeal has been mentioned by Julka and Senapati (1987) and Guerro (1981, 1983) respectively. A small white earthworm, *Enchytraeus albidus,* usually grown in the soil, is very useful for aquarium fish and small laboratory animals. Tacon *et al.* (1983) considered earthworms to have good potential for complete protein supplementary food for *Tilapia* fish growth. Jin-you *et al.* (1982) demonstrated faster weight gain in chicken, fed on earthworms. They have also reported that piglets reared on a diet containing vermiprotein than other protein supplements, showed accelerated weaning, earlier estrus in sows, better disease tolerance and declining the cases of diarrhoea. According to Kale and Bano (1984) the protein content of worm meal is considerably greater than that of the fishmeal and soybean meal proteins. Enhanced growth rates were observed when earthworms were fed to chicks crabs, fishes and tadpoles. Substitution of fishmeal in the poultry feed by worm meal was reported to have no difference in the body weight of the broilers (Kale, 1985). Arunachalam and Palanichamy (1984) have reported improved growth rate of fish *Mystus vittatus* when fed on dry or fresh worms. Worm meal could be a suitable supplement / replacement of traditional food. It can be prescribed for protein deficiency in human. Dried earthworm powder has been demonstrated to have preventive and curative properties of a variety of diseases.

Earthworms as source of medicines

Earthworms have been used to cure various human diseases since ancient times, for more than 4000 years (2600 B.C.), in various parts of the world including China, Japan, Korea, India, Cambodia, Myanmar (Burma), Vietnam, Iran and Middle East. Both external application and internal intake of such medicines have been suggested. The ointments and extracts prepared from earthworm tissues have been used for the treatment of numerous diseases. In Ayurveda and Unani systems of medicine (India) the paste of dried worms has been recommended for treatment of wounds, chronic boils, piles, chronic cough, sore throat, teeth diseases, hernia, impotency, small pox, smooth delivery, gall bladder stone, hair growth, diphtheria, jaundice, fevers, rheumatic pains, tuberculosis, bronchitis, facial paralysis, scorpion and snake bite etc. In India, some traditional physicians (Vaidyas) prepare medicines using ashes and extracts (Bhasma and Kalpa) of earthworms.

In China, earthworms have been used as drugs under the name **"Jiryu"** for thousands of years. Powder of dried earthworms have been prescribed for the treatment of fevers, particularly related with lungs, as an antidote for poisons, for breathing difficulties, coughs and for excess water retention. In ancient and famous Chinese medical text **"Ben Cao Gang Mu"** (Compendium of Medicine), earthworms **(Di-Lung)** are described as "salty in taste, cold in property, efficacious in clearing the heart, invigorating blood circulation, dissolving stasis, opening up channels, curing stroke, hemiplegia and infantile convulsion". In Burma, worm medicines have been used as folk medicine for treatment of pyorrhea, small pox and milk enhancer to in nursing mothers. In Iran earthworm drugs have been used for hair growth and expulsion of gall bladder stones (Stephenson, 1930). According to historical records earthworms are useful in curing rheumatism (Reynolds and Reynolds, 1972). In the Middle East paste of earthworm ashes in rose oil is used as hair growth formula. In Japan four kinds of drugs are prepared from earthworms, antibiotics, aphrodisiacs, antipyretics and antidotes.

Several workers have described the medicinal properties of earthworms (Kellin, 1920; Stephenson, 1930; Weibach, 1962; Reynolds and Reynolds, 1972; Vohra and Khan, 1978; Mishra and Dash, 1980; Polunin and Robbins, 1992; Cooper and Wang, 2004; Chaudhuri, 2005; Costa-Netto, 2006; Ranganathan, 2006). However, only few scientific investigations have been made inclusive of double blind clinical trials.

Charles Darwin has mentioned about the ability of earthworms in dissolving and digesting complex materials. Strong activities of digestive enzymes, particularly the serine proteases have later been demonstrated in the alimentary canal of the earthworms.

Impressed by the use of earthworm powders in traditional Chinese medicine (TCM) for cardiovascular disorders, earthworm enzyme preparations were tested for fibrinolytic activity. During 1980s, a strong fibrinolytic enzyme complex was isolated and purified from earthworm, *Lumbricus rubellus* (Mihara *et al.*, 1983, 1989, 1991, 1992, 1993; Lu *et al.*, 1988; Nakajima *et al.*, 1993, 2000; Lin *et al.*, 2000; Cho *et al.*, 2004). Earthworm fibrinolytic enzyme (EFE) was demonstrated to include six proteolytic enzymes collectively named as lumbrokinase (LK). The enzyme (s) were reported to be fibrin specific, showing high activity in the presence or absence of plasminogen and are potentially very useful in treating thrombosis and dissolving intra vascular fibrin clots. As compared to other thrombolytic drugs, EFE is cheaper, easy to store (high stability) and effective through oral route (Mihara *et al.*, 1991; Hahn *et al.*, 1997; Nakajima *et al.*, 2000; Kim *et al.*, 1998). Therapeutic and preventive effects, of the preparation, for thrombosis related diseases have been confirmed clinically (Jin *et al.*, 2000). It was suggested that LKs might be used for prevention of cerebral infarction and in patients with a previous cerebrovascular ischemic event.

Successful attempts have been made to isolate, purify, characterize, crystallize EFE from other species of earthworms, *Eisenia fetida* (Yang and Ru, 1997; Wang *et al.*, 2003), *E. andrei* (Lee *et al.*, 2007). Instead of 6 enzymatic components, 7 have been found in these EFE and some differences in isoelectric points and their hydrolytic activity have also been demonstrated. Fibrinolytic activity has also been found in *Eudrilus eugeniae* (Agrawal *et al.*, Unpublished data).

In 1983, Dr. Hisashi Mihara presented his findings at the International Society and Thrombosis and Hemostasis Conference in Stockholm. In 1997, Chinese government approved Plasmin (earthworm enzyme complex) as a new medicine. Subsequently, it was approved by various organizations and bodies in China and in other countries. Now product has been patented in several (23) countries under different trade names like Plasmin Plus, vermivit, Bolouake, LR-Zyme and are recommended to be used as a dietary supplement for those who require to improve their blood circulation, especially those who suffer from cardiovascular difficulties, both acute and chronic, such as cerebral thrombus, cerebral embolism, pulmonary embolism, angitis, myocardial infarction, hyperplasminemia, platelet hypercoagulability, deep vein thrombosis, traveler's thrombosis, blood clots, tendency to develop thrombus. It is nontoxic, good for long term use without any side effects and helps to maintain a healthy balance between hemolysis and hemostasis, improves microcirculation, repairs damaged nerve cells and reduce blood sugar. This is an excellent example of successful development of an effective drug formulation with state of the art of bioengineering technology. The product (LR-Zyme) is designed and formulated to be given to pet animals (Fig. 3).

Fig. 3: Earthworm based drugs Plasmin Plus and LR- Zyme

Hori *et al* (1974) have reported significant anti-pyretic activity in the tissue extracts of earthworms, *Lumbricus spencer* and *Perichaeta communissuina* in experimental rabbits and it was concluded that the effect might be due to monoamines. Nagasawa *et al* (1991) have demonstrated inhibition of the growth of mammary tumours in mice by lombricine obtained from earthworms. The findings also indicate that the drug may also be of some importance for diabetic patients. Anti-tumor and anti-cancer properties of EFE and some other earthworm based bioactive molecules have been demonstrated using experimental animals and culture cell lines (Chen, 2001; Lee *et al.*, 2005, 2007). Further studies are still required in this direction.

In some regions of Southern India, earthworm paste is commonly used in acute inflammatory conditions. Vohra and Khan (1978) described that in Unani medicine, pulverized worms are used in the treatment of cellulites, boils, rheumatism and other inflammatory conditions. Significant anti-inflammatory of crude and fractioned earthworm extracts has clearly been demonstrated in experimental rats in which paw oedema was induced by carageenan injection and cotton pellets implantation (Yegnarayanan *et al.*, 1987, 1998; Ismail *et al.*, 1992; Balamurugan *et al.*, 2005). Coelomic fluid of earthworms is an interesting and highly active fluid mixture that seems to be the source of many bioactive substances such as lectins, antimicrobial peptides (fetidins), cytolytic and pore forming proteins (eiseniopore, lysenin, cytolistin), proteases and phenoloxidases etc.). A glycolipoprotein mixture (G-90) has been isolated from tissue extracts of *Eisenia fetida*. These are responsible for a variety of biological activities namely, hemolytic, agglutinating, cytolytic, cytotoxic, bacteriostatic, anticoagulating, thrombolytic, toxic, anti-tumour, antioxidant and antimicrobial properties. Therapeutic significance of these materials are being explored in various laboratories (Roach *et al.*, 1984; Vailliger *et al.*, 1985; Kobayashi *et al.*, 2001; Popovic *et al.*, 2005).

Conclusion

Earthworms can do wonders for us, help us to solve the problems of waste management; revitalize the soils; increase crop production without disturbing the environmental quality; boost up crop quality (organic farming and sustainable agriculture) and community (human) health by improving nature more natural and providing a renewable source of health promoting products, vermimeal, vermiprotein, vermiceuticals. Awareness programmes should be launched at the village and block levels by the scientific communities, NGOs, Universities to popularize the knowledge about the importance and utility of earthworms and their role in solid waste management. Rural, unemployed youths should be trained properly for vermicomposting techniques and employment generation and about the earthworm based products too.

References

Agrawal, Dheeraj and Agrawal, O.P. 2006a: Hanging Vermicomposting Units: A Novel Approach. VII South Asian Symposium on Odonatology and Recent Trends in Zoology, Hislop College, Nagpur.

Agrawal, O.P. 2005a: Vermicomposting practice-conversion of garbage into gold. *STTPP WAMR*, p.188-196.

Agrawal, O.P. 2005b: Vermitechnology for all-round sustainable development. National Seminar on Composting and Vermi-composting held at CSRTI, Mysore, October 26, 27. p. 39-47.

Agrawal, O.P. and Dheeraj Agrawal, 2006: Vermitechnology can take care of our garbage and can solve many of our problems. Reading Manual for "Vacation training Programme on Bioresources for School Children", 15 May - 15 June. School of Studies in Zoology, Jiwaji University, Gwalior, p. 117-128.

Arunachalam, S. and Palanichany, S. 1984: Earthworms as a feed for the catfish, *Mystus vittatus*. *National Seminar on Organic Waste Utilization and Vermicomposting*. School of Life Sciences, Sambalpur University, Orissa, p. 4.

Balamurugan, M., Prakash, M., Parthasarathi, K., Cooper, E.I. and Ranganathan, L.S. 2005: Effect of earthworm paste (*Lampito mauritii*, Kinberg) on the anti-inflammatory, anti-oxidative, haemastological and serum biochemical indices of rat (*Rattus norvegicus*). *Evid. Based Complement Altern. Med.* **2:** 1-3.

Barrett, T.J. 1959: *Harnessing the Earthworms*. Faber and Faber, London. p.166.

Bouche, M.B. 1977: Strategies lombricienns. *Ecol. Bull.* (Stockholm). **25**: 122-132.

Chaudhuri, P.S. 2005: Vermiculture and Vermicomposting as Biotechnology for conversion of organic wastes into animal protein and organic fertilizer. *Asian J. Microbiol. Biotech. Env. Sci.* **7:** 359-370.

Chen, H. 2001: Anti-tumor effects of earthworm extracts EF2. *Chin. Clin. Oncol.* (Chin). **6**: 349-350.

Chen, H. 2007: Earthworm fibrinolytic enzyme: anti-tumor activity on human hepatoma cells *in vitro* and *in vivo*. *Clin. Med. J.* **120**: 898-904.

Cho, I.H., Choi, E.S., Lim, H.G. and Lee, H.H. 2004: Purification and characterization of six fibrinolytic serine-proteases from earthworm *Lumbricus rubellus*. *J. Biochem. Mol. Biol.***37**: 199-205.

Cooper, E.L and Wang, Ru B. 2004: Earthworms: sources of antimicrobial and anticancer molecules. *Adv. Exp. Med. Biol.* **546:** 359-389.

Costa-Netto, E.M. 2005: Animal-based medicines: biological prospection and the sustainable use of zootherapeutic resources. *An. Acad. Brass. Cienc.* 77: 33-43.

Darwin, C. 1881: *The Formation of Vegetable Mould Through the Action of Worms, With Observation of Their Habits*. Murray, London.

Edwards, C.A. 1985: Production of feed protein from animal wastes by earthworms. *Phil. Trans. Res.Soc, London*, **310**: 299-307.

Edwards, C.A. and Bohlen, P.J. 1996: *Biology and Ecology of Earthworms*. Chapman and Hall, London.

Edwards, C.A. and Lofty, J.R. 1977: *Biology of Earthworms*. Chapman and Hall, London.

Edwards, C.A. and Niederer, A. 1988: The production and processing of earthworm protein: In: *Earthworms in Waste and Environment Management*. (Eds. Edwards, C.A. and Neuhauser, E.F.). SPB Academic Publishing, The Netherlands, p. 169-180.

Eijasackers, H. 1998: Earthworm in environment research: Still a promising tool. In: *Earthworm Ecology* (Ed. Edwards, C.A.), CRC Press, The Netherlands, p. 295-323.

Ghatnekar, S.D., Kavian, M.F., Ghatnekar, G.S. and Ghatnekar, M.S. 1998: Management of solid wastes through vermiculture biotechnology. In: *Ecotechnology for Pollution Control and Environmental Management* (Eds. Trivedi, R.K. and Arvind Kumar). Environ Media, Maharastra, p. 59-67.

Guerrero, R.D. 1981: The culture and use of *Perionyx excavatus* as a protein resource in Philippines. In: *Proceedings of Darwin Centenary Symposium.* Grange-over-sanda, Great Britain, p. 22.

Guerrero, R.D. 1983: The culture and use of *Perionyx excavatus* as protein resources in the Philippines. In: *Earthworm Ecology from Darwin to Vermiculture* (Ed. Satchell, J.E.). Chapman & Hall, London, p. 309-313.

Hahn, B.S., Jo, Y.Y., Yang, K.Y., Wu, S.J., Pyo, M.K., Yun-Choi, H.S. and Kim, Y.S. 1997: Evaluation of the *in vivo* antithrombotic, anticoagulant and fibrinolytic activities of *Lumbricus rubellus* earthworm powder. *Arch. Pharm. Res.* **20**: 1-23.

Hori, M., Kondon, K., Yoshida, T., Konishi, E. and Minami, S. 1974: Studies of antipyretic components in the Japanese earthworm. *Biochem. Pharmacol.* **23:** 1583-1590.

Ismail, S.A., Pulanmdiran, K. and Yegnarayan, R. 1992: Anti-inflammatory activity of earthworm extracts. *Soil Biol. Biochem.* **24:** 1253-1254.

Ismail, Sultan A. 1997: *Vermicology: Biology of Earthworms.* Orient Longman, Publication, Hyderabad.

Jin, L., Jin, H., Zhang, G. and Xu, G. 2000: Changes in coagulation and tissue plasminogen activator after the treatment of cerebral infarction with lumbrokinase. *Clin. Hemorheol. Microcirc.* **23:** 213-218.

Jin-you, X., Xian-kuan, Z. and Zhie-ren, P. 1982: Experimental research on the substitution of earthworm for fishmeal in feeding broilers. *J. South China Normal College.* **1**: 88-94.

Julka, J.M. 1988: *The Fauna of India and the Adjacent Countries. Megadrile:Oligochaetidae.* ZSI, Calcutta, India.

Julka, J.M. and Senapati, B.K. 1987: *Records of the Zoological Survey of India.* Miscellaneous Publications, Grafic Printall, Calcutta, India, p. 1-45.

Kale, R.D. 1998: *Earthworm: Cinderella of Organic Farming.* Prism Books Pvt. Ltd., Bangalore.

Kale, R.D. and Bano, K. 1984: Earthworm cultivation and culturing technique. *National Seminar on Organic Waste Utilization and Vermicomposting.* School of Life Sciences, Sambalpur University, Orissa, p.7.

Kellin, D. 1920: On the pharyngeal or salivary gland of the earthworm. *Q. J. Microsc. Sci.,* **65:** 33-61.

Kim, Y.S., Pyo, M.K., Park, K.M., Hahn, B.S., Yang, K.Y. and Yun-Choi, H.S. and 1998: Dose dependency earthworm powder on antithrombotic and fibrinolytic effects. *Arch. Pharm. Res.* **21**: 374-377.

Kobayashi, H., Ohtomi, M., Sekizawa, Y. and Ohta, N. 2001: Toxicity of coelomic fluid of the earthworm *Eisenia fetida* to vertebrates but not invertebrates: probable role of sphingomyelin. *Comp.Biochem. Physiol.* C **128:** 401-411.

Lawrence, R.D. and Miller, R.H. 1945: Protein contents of earthworms. *Nature* **.39**: 517.

Lee, K.C., Shin, J.S., Kim, B.S., Cho, I.H., Kim, Y.S and Lee, E.B. 2007: Antithrombotic effects by oral administration of novel proteinase fraction from earthworm *Eisenia andrei* in venomous thrombosis model in rats. *Arch. Pharm. Res.* **30**: 475-480.

Lee, K.E. 1985: *Earthworm: their Ecology and Relationship with Soil and Land Use.* Academic Press, New York.

Lin, S.Q., Yu, P. and Lan, R.F., 2000: Application of affinity chromatography for purification of fibrinolytic enzymes from earthworm. *Pharm. Biotech. (Chin.).* **7:** 229-233.

Linnaeus, C. 1758: *Systema Naturae.* Regnum Animale. **10:** 824.

Lu, Y.H., Jin, R.C. and Wu, Y.W. 1988: Separation and purification of fibrinolytic enzymes from earthworm. *J. Biochem. (Chin.).* **4:** 166-172.

Maity, S., Roy, S., Chaudhury and Bhattacharya 2007: Antioxidant responses of the earthworm *Lampito mauritii* exposed to Pb and Zn contaminated soil. *Environmental Pollution* **10:** 1-7.

Mihara, H., Maruyama, M. and Sumi, H. 1992: Novel thrombolytic therapy discovered from traditional Oriental medicine using the earthworm. *SE Asian J. Trop. Med. Public Health.* 23 Suppl **2:** 131-140.

Mihara, H., Nakajima, N. and Sumi, H. 1993: Characterization of protein fibrinolytic enzyme in earthworm, *Lumbricus rubellus. Biosci. Biotech. Biochem.* **57**: 1726-1731.

Mihara, H., Sumi, Akazawa, T., Yineda, T. and Majumoto, H. 1983: Fibrinolytic enzyme extracted from the earthworm. *Throm Haemostas.* **50:** 258-263.

Mihara, H., Sumi, H., Yoneta, T., Mizumoto, H., Ikeda, R., Seikl, M. and Maruyama, M. 1991: A novel fibrinolytic enzyme extracted from the earthworm, *Lumbricus rubellus. Jpn. J. Physiol.* **41:** 461-472.

Mihara, H., Yoneta, T., Sumi, H., Soeda, M. and Maruyama, M. 1989: A possibility of earthworm powder as therapeutic agent for thrombosis. *Thromb. Haemosta.* **62:** 545-549.

Mishra, P.C. and Dash, M.C. 1980: Digestive enzymes of some earthworms. *Experentia,* **10:** 1156-1157.

Nagasawa, H., Sawaki, K., Fujii, Y., Kobayashi, M., Segawa, T., Suzuki, R. and Inatomi, H. 1991: Inhibition by lombricine from earthworm (*Lumbricus terrestris*) on the growth of spontaneous mammary tumours in SHN mice. *Anticancer Res.* **11:** 1061-1064.

Nakajima, N., Mihara, H. and Sumi, H. 1993: Characterization of potent fibrinolytic enzymes in earthworm, *Lumbricus rubellus. Biosci. Biotech. Biochem.* **57**: 1726-1730.

Nakajima, N., Sugimoto, M. and Ishihara, K. 2000: Stable earthworm serine proteases: Application of the protease function and usefulness of the earthworm autolysate. *J. Biosci. Bioeng.* **90:** 174-179.

Polunin, M. and Robbins, C. 1992: *The Natural Pharmacy: An Illustrative Guide to Natural Medicine.* Collier Books, McMillan Publishing Company, New York.

Popovic, M., Grdisa, M. and Hrzenjak, T. 2005: Glycoprotein (G-90) obtained from the earthworm, *Eisenia fetida* exerts antibacterial activity. *Veterinarski Arhiv.* **75:** 119-128.

Popovic, M., Hrzenjak, T., Babic, T., Kos, J. and Grdisa, M. 2001: Effect of earthworm (G-90) extract on formation and lysis of clots originated from venous blood of dogs with cardiopathies and malignant tumors. *Pathol. Oncol Res.* **7:** 197.

Ranganathan, L.S. 2006: *Vermibiotechnology: From Soil Health to Human Health.* Agribios (India), Jodhpur.

Reynolds, J.W. and Renolds, W.M. 1972: Earthworms in medicine. *Am. J. Nursing.* **72**: 1273.

Roch, P., Davant, N. and Lassegnes, M. 1984: Isolation of agglutionins from lysis in earthworm coelomic fluid by gel filtration than chromatofocussing. *J. Chromat.* **290:** 231-235.

Roy, P. 2003: Treatment of municipal solid waste. *Everyman's Science.* **38**: 187-190.

Satchell J.E. 1967: Lumbricidae. In: *Soil Biology.* (Eds. Burges, A. and Raw, F.), Academic Press, London. p. 259-322.

Satchell J.E. 1983: Earthworm ecology from Darwin to Vermiculture, Chapman and Hall, London. p.495.

Singh,J. and Rai, S.N. 1998: Potential of earthworms in sustainable agriculture. *Yaajna.* November p.10-12.

Stephenson, J. 1930: *The Oligochaeta.* Clerendon Press, Oxford.

Tacon, A.G.J., Stafford, E.A. and Edwards, C.A. 1983: A preliminary investigation of the nutritive value of three terrestrial worms for rainbow trout. *Aquaculture* **35**: 187-199.

Talashikar, S.C. and Powar, A.G. 1998: Vermitechnology for ecofriendly disposal of waste. In: *Ecotechnology for Pollution Control and Environmental Management.* (Eds. Trivedy, R.K. and Kumar, A.), Global Science Publications, Karad, p. 171-197.

Valliger, J., Cadoret, M.A., Roach, P. and Valembois, P. 1985: Protein analysis of earthworm coelomic fluid. III. Isolation characterization of several bacteriostatic molecules from *Eisenia fetida andrei. Develop. Comp. Immunol.* **9:** 11-20.

Vohra, S.B. and Khan, M.S.Y. 1978: *Animal Origin of Drugs Used in Unani Medicine.* Institute of History of Medicine and Medical research, Tughlaquabad, New Delhi.

Wang, F., Wang, C., Li, M., Gui, L., Zhang, J. and Chang, W. 2003: Purification, characterization of a group of earthworm fibrinolytic enzymes from *Eisenia fetida. Biotechnol. Letters* **25**: 1105-1109.

Weibach, W.W. 1962: Regenwurmer und Essbare Erde. *Biol. Jaarb. Dodonea.* **30:** 225-238.

Yang, J.S. and Ru, B.G. 1997: Purification and characterization of an SDS-activated fibrinolytic enzyme from *Eisenia fetida. Comp. Biochem. Physiol.* **118B**: 623-631.

Yegnarayanan, R., Ismail, S.A. and Shorti, D.S. 1998: Anti-inflammatory activity of two earthworm portions in carrageenan pedal oedema test in rats. *Indian J. Physiol. Pharmacol.* **32:** 72-74.

Yegnarayanan, R., Sethi, P.P., Rajhans, P.K., Pulandiran, K. and Ismail, S.A. 1987: Anti-inflammatory activity of total earthworm extracts in rats. *Indian J. Pharmac.* **19:** 221-224.

Chapter **6**

Therapeutic Values of Earthworm Paste

L.S. Ranganathan*, M.Balamurugan and K.Parthasarathi
**V.L.B. Janakiammal College of Arts and Science, Kovaipudur, Coimbatore- 641 042, India.*
Division of Vermibiotechnology, Department of Zoology, Annamalai University, Annamalainagar 608002, India.

Zoo therapy - the science of curing human ailments by using therapeutics obtained from animals is an ancient and traditional medical technology. Animal-based medicines have been prepared from parts of the animal body, their secretory and excretory products, or from non-animal materials (nests and cocoons). Studies on traditional wisdom on medicines have given a vital tool in the upcoming art of bio-prospecting for pharmaceutical compounds. Of the 252 essential compounds selected by the World Health Organization for medicinal purposes, 11.1% are derived from plants and 8.7% from animals (Costa-Neto, 2005).

Earthworms, apart from being used as food for man (by Maoris in New Zealand, Japanese, natives in New Guinea, Africa) and animals, because of their high protein content, (64 - 72% of dry weight) are eaten or applied to human to cure illness or diseases such as piles, fever, small pox, jaundice and removal of stones in bladder. Apart from proteins, earthworm tissues contain essential aminoacids like phenylalanine, leucine, lysine, isoleucine, methionine and trace elements, vitamins, enzymes and antioxidants (Ranganathan, 2006).

For thousands of years, earthworm and its product had been used for its therapeutic benefits. The traditional medical knowledge of indigenous people throughout the world, more particularly in Asia, including India, Myanmar, China, Korea and Vietnam has played vital role in identifying, extracting and using biologically active compounds from earthworms. For almost 4,000 years, China has a history in research on medicinal use of earthworms. Earthworm, formulated as "earth dragon", has been used as a suitable drug for the gastro-intestinal disorder, particularly inflammation caused in gastrointestinal tract. "Earthworm tonic" is also used for balancing the sympathetic and parasympathetic functions of the central nervous system. In Korea taking "earthworm soup" before going to bed is a traditional habit, which is believed to enrich general health and prevent various ailments. Vietnamese were known to use earthworms as a "Miracle Medicine that could save lives in 60 minutes" for acute multiple organ dysfunction, blood infection, hemorrhagic fevers, severe burns and strokes. Earthworm has a strong taste-"salty", "fishy" or both; it may result in nausea and even vomiting in some sensitive individuals. Taking earthworm powder or taking earthworm with citrus fruits or other herbs can counter this nausea and vomiting. 15 to 20% of ayurvedic medicines in India are based on animal derived substances. "Charaka samhita" - the oldest available ayurvedic classic refers to 380 types of animal substances. 'Vermivit' is an alcoholic extract from *Eisenia fetida,* which is used in the treatment of numerous diseases. 'Di-Long' – the earthworm powder has been shown to have analgesic, antipyretic, blood pressure reducer, anti-coagulative, wound healing and anti cancer effects, G-90- a glycoprotein derived from *E.fetida* had been shown to have lectine activity, acts as agglutinin, anticoagulant, fibrinolytic agent and antioxidant effect. Till now, there are no known drug interactions with earthworms. Taking earthworm medicine with any other herbal remedies or dietary supplements is practiced by tribal peoples. Earthworm and its derivatives were known to be used for various ailments in different parts of India, particularly by tribes in Western Ghat and Tripura. Various studies on earthworms have shown it to exhibit antipyretic, antispasmodic, detoxic, diuretic, antihypertensive, antiallergic, antiasthmatic, spermatocidal, antioxidative, antimicrobial, anticancer, antiulceral and anti-inflammatory activities. Earthworm derived medicines were claimed to cure fever, relieve pains, blood pressure, promote fertility, increase immunity, heal wounds, arthritis, asthma, inflammation, arteriosclerosis and aging. Because of these important properties and its use in Chinese traditional medicines, earthworms are expected to play important role in future clinical practices. The present study describes the therapeutic (anti-ulceral, anti-inflammatory and anti-oxidative) efficacy of earthworm paste prepared from *Lampito mauritii* (Kinberg).

Preparation of "earthworm paste"(EP)

Earthworms, *Lampito mauritii* (Kinberg) were obtained from the stock culture of the Division of Vermibiotechnology, Department of Zoology, Annamalai University. Five hundred sexually matured, clitellated worms (900 mg/worm) were washed with running tap water and then fed with wet blotting paper for a day to clear the gut. These worms were again washed with distilled water. They were kept in plastic trough covered tightly with polythene cover and exposed to sunlight for three days to kill them. Mucus and coelomic fluid that oozed out digested the dead worms forming a brown coloured paste referred to as "earthworm paste" (EP) (Balamurugan et al., 2007).

Animal model

Healthy and pure strain male albino rats (*Rattus norvegicus*), (150-200 g body weight) were procured from the Department of Experimental Medicine, Central Animal House, Rajah Muthiah Medical College, Annamalai University and used for the experimental study. The animals were housed in polypropylene cages at 24° ± 2° C and were fed with proper food and water ad libitum throughout the experiment. The experiment got clearance from the institutional animal ethical committee (IAEC).

Aspirin plus pyloric ligation induced ulcer model (Goel *et al.* 1986 and Parmar *et al.*1984).

The rats for anti-ulceral study were divided into 8 groups of 6 animals each. Group I (normal control) received water. Group II (aspirin control) received aspirin (200 mg/kg). The remaining 6 groups served as experimental groups receiving ranitidine (50 mg/kg) - a standard drug and "earthworm paste" administered in different doses (20, 40, 80 and 160 mg/kg). All the doses were administered orally, once daily for 10 days. From the 6th day onwards, animals in groups III to VIII received aspirin (200 mg/kg) orally, one hour after the administration of the drugs. On the 11th day pylorus ligation was carried out on the 18 hours fasted rats. After 4 hours of pylorus ligation, the animals were sacrificed by decapitation. The stomach was cut open along the greater curvature and the gastric juice was collected and centrifuged at 3000 rpm for 10 minutes. The supernatant was measured and used for the estimation of total and free acidity. The stomach was washed with normal saline and the lesions were measured by adopting a scale and the ulcer index was determined where loss of normal morphology was given 1 point, discolouration of mucosa 2 point, ulcer up to 1 mm (dia.) 3 point: ulcer up to 2 mm (dia.) = 4 points

Estimation of free and total acidity

1 ml supernatant was pipetted out and titrated against 0.01 N Sodium hydroxide using Topfers reagent and Phenolphthalein as an indicator.

Carageenan-induced rat paw edema: Acute model study

The rats (either sex) were divided in to seven groups comprising six animals in each. Of these seven groups, control animals received only 2% gum acacia, second group received aspirin (75 mg/kg) and other five experimental groups received EP orally in different doses (20, 40, 80 and 160 mg/kg). Aspirin and EP were suspended in 2% gum acacia and administered orally one hour prior to the sub plantar injection of 1% carageenan (1 ml/100 g/body weight). One hour after the drug administration the paw edema (Winter et al., 1962) volume was measured by using mercury plethysmograph at 0 and 3 hrs. Mean increase in paw volume was measured and percentage inhibition calculated.

Granuloma pouch method: Chronic model study

Subcutaneous dorsal granuloma pouch was induced in ether-anaesthetized rats by injecting 25 ml of air, followed by injection of 0.5 ml of turpentine oil (Selye, 1953). All drugs were administered orally one hour prior to turpentine oil injection and continued for seven consecutive days. On day 7th, the pouch was weighed and amount of exudates measured and compared with those of the control and standard group.

Determination of antioxidant activity

The total antioxidant activity was determined according to the method described by Mitsuda et al., (1996) with some modifications. A 1 ml aliquot of linoleic acid emulsion (2 mg/ml in 95% methanol) was mixed with 1ml sample solution (0.1 mg/ml) and 0.2 ml of phosphate buffer (pH 7.0, 0.2mol/l). After incubation at 37 ± 1°C in the dark for 72 h, a 0.1 ml aliquot of reaction solution was mixed with 4.7 ml of ethanol (75%), 0.1 ml of ammonium thiocyanate (30%) and 0.1 ml of ferrous chloride (20 m mol/l). After the mixture was stirred for 3 min, the peroxide value was determined by reading the absorbance at 500 nm, and the inhibition percent of linoleic acid per oxidation was calculated as : (%) inhibition= [1-(absorbance of sample at 500 nm)/(absorbance of control at 500 nm)]×100. Ascorbic acid (0.1mg/ml) was used as positive control.

DPPH radical scavenging activity assay

Free radical scavenging activity against 2,2-diphenyl-1-picrylhydrazyl (DPPH) radical was measured using the method of Oboh (2005) with modifications. Different dilutions of the extracts were prepared in triplicate. Then 1ml of each dilution was added to 2 ml of 0.15 mM of DPPH. The mixture was shaken vigorously and allowed to stand at room temperature in the dark for 30 min. and the absorbance was measured at 517 nm in a spectrophotometer.

Lower absorbance of the reaction mixture indicated higher free-radical scavenging activity. The scavenging activity of sample was expressed as 50% effective concentration (EC-50), which represented the concentration of sample having 50% of DPPH radical scavenging effect. Antioxidant activity (AA) was expressed as the percentage of DPPH decrease using the equation:

$AA (\%) = [1-(A_{control} - A_{sample})/A_{control}] \times 100.$

EC-50 of the extract was determined from the graph of antioxidant activity (%) against amount of extract (mg). Ascorbic acid was used as a standard.

EXPERIMENTAL FINDINGS

Study of anti-ulcer activity

Aspirin, an ulcerogen and pyloric ligation had significantly increased the gastric juice volume (6.58 ± 0.43ml), free acidity (28.2 ± 0.12) and total acidity (55.3 ± 0. 51) except pH compared to normal control. These symptoms of ulcer were brought to near normal condition when the standard drug-ranitidine was administered." Earthworm paste" administration, particularly 160 mg/kg, had significantly decreased the gastric juice volume 4.1 ± 0.29 ml, free acidity 11.1 ± 0.15 and total acidity 24 ± 0.28 in the ulcerated rats. And these values were better than treatment with ranitidine. The ulcer index in aspirin treated rat was 17.05 ± 0.17. This was found to get reduced in the "earthworm paste" treated animal. Particularly 160 mg of "earthworm paste"/kg treated rats showed better reduction in ulcer index (07.50 ± 0.31) (Table 1).

Table 1. Anti-ulceral activity of earthworm paste on stomach of *Rattus norvegicus* ($P < 0.05$).

Treatments	Gastric juice volume (ml)	pH of Gastric Juice	Free acidity	Total acidity	Ulcer index
Normal (control)	6.43 ± 0.37	6.4 ± 0.18	09.2 ± 0.18	17.29 ± 0.27	-
Ulcerated rat (aspirin)	6.58 ± 0.43	3.4 ± 0.15	28.2 ± 0.12	55.3 ± 0. 51	17.05 ± 0.17
20 mg/kg	6.3 ± 0.20	4.4 ± 0.12	16.4 ± 0.15	35.27 ± 0.21	12.15 ± 0.30
40 mg/kg	6.8 ± 0.18	4.3 ± 0.17	14.2 ± 0.10	32.40 ± 0.24	10.25 ± 0.34
80 mg/kg	6.1 ± 0.68	5.5 ± 0.18	11.1 ± 0.15	24.28 ± 0.21	09.25 ± 0.11
160 mg/kg	4.1 ± 0.29	5.9 ± 0.12	10.5 ± 0.19	22.18 ± 0.21	07.50 ± 0.31

Anti-Inflammatory Status

It is observed in the present study that the carageenan induced acute phase rat hind paw edema volume and the turpentine induced chronic phase granuloma pouch weight and the volume of fluid were reduced significantly due to the administration of aspirin; whereas the administration of EP was found to exhibit better results. Administration of 80 mg/kg EP was found to reduce all the above parameters and brought the values to near normalcy, and this was found after administration of 40, 20, 160 and 320 mg/kg, respectively (Table 2).

Table 2. Anti-inflammatory activity of earthworm paste on inflammed *Rattus norvegicus* ($P < 0.05$).

Treatments	Volume of Paw edema (ml)	Granuloma Weight (g)
Inflammed rat	2.37 ± 0.04	3.18 ± 0.04
20 mg/kg	0.64 ± 0.05	1.72 ± 0.34
40 mg/kg	0.19 ± 0.13	0.10 ± 0.16
80 mg/kg	0.66 ± 0.26	0.15 ± 0.43
160 mg/kg	0.49 ± 0.03	0.77 ± 0.54
320 mg/kg	0.10 ± 0.45	0.52 ± 0.18

Total antioxidant capacity and free radical scavenging power

Peroxide value was determined by reading the absorbance at 500 nm. The percentage inhibition of per oxidation in linoleic acid emulsion for earthworm paste was 32.3 %. The antioxidant efficacy by the earthworm extract was very significant (Table 3).

Table 3. Total anti-oxidant capacity and DPPH scavenging activity of earthworm paste (P<0.05)

Antioxidant activity (%)	DPPH scavenging activity EC-50 (μg/ml)
32.3±1.2	91±1.4

DPPH is a stable free radical and accepts an electron or hydrogen radical to become a stable diamagnetic molecule. The earthworm paste exhibited potent antioxidant activity, efficiently scavenging the DPPH free radical with an EC50 value of 91 μg/ml (Table 3).

Ulcer is due to the imbalance between two opposing factors: (a) attacking factors including - *Helicobacter pylori*, bile salts, acid and pepsin and (b) defensive factors including - mucus and gastric mucosal barrier particularly cyclo-oxygenase which is responsible for mucus production, improved blood flow, and bicarbonate production in the duodenum. Attacking factors predominate in duodenal ulcer and failed defensive mechanism in gastric ulcer. Consequently the treatment of peptic ulcer is directed against either reduction of the aggressive factors or enhancement of defensive mechanism.

Ulcerogens like ACTH, cortisone, aspirin and phenylbutazone reduce the rate of secretion of mucus and reduce the concentration of protein bound carbohydrates in these secretions (Bockus, 1963). Aspirin, a non-steroidal anti-inflammatory drug, induces gastric ulcer by causing back diffusion of H+ ions into the mucosal cells (Davenport 1969), reduce mucosal defense through inhibition of cyclo-oxygenase (COX) and prostaglandin synthesis and is also known to increase the gastric juice secretion, total acidity, free acidity and reduce the pH. Davenport (1969) reported that the aspirin-induced damage to the gastric mucosa is associated with breakage of gastric mucosal barrier. Aspirin induced ulcer is used as experimental evaluation process of anti-ulceral activity. In the present study, administration of "earthworm paste", particularly 160 mg/kg was found to bring down the gastric juice volume, free acidity, total acidity and gastric ulcer index clearly indicating its efficacy in reducing these factors, which lead to ulcer. These results were even better than the results of the standard drug - ranitidine - used for treatment of ulcer. Various mechanisms of actions of anti-ulceral drugs which reduce the gastric secretion have been shown: (i) Captopril contributing SH group (Roa et.al., 1995), (ii) Levcromakalim and Nicorandil opening potassium channel (Bose et al 2003) and (iii) polyphenolic compounds inhibiting histamine H2 receptor (Martin et al., 1988). The reduction in the gastric secretion of "earthworm paste" treated rats could be due to the highly proteinaceous nature of earthworm, particularly SH containing aminoacids and due to the presence of polyphenol in the tissues of earthworms which could inhibit histamine receptor (Ranganathan, 2006). Selected flavonoids have been shown to have antiallergic, anti-inflammatory, antiviral and antioxidant activities (Middleton, 1996). Flavonoids, a polyphenol, were known to exert their anti-ulcer activity by protecting the mucosa by preventing formation of lesions (Ebadi, 2003). "Earthworm paste" by in the present study, was shown to reduce lesions reflected by low ulcer index. Since significant amount of polyphenolic compounds are known to be present in earthworm tissues (Ranganathan, 2006) it is likely that these polyphenolic compounds were responsible for the inhibition of histamine receptor and reduction of gastric juice and thereby to the protection of gastric mucosa.

Carageenan-induced rat paw edema is commonly used in evaluating the anti-inflammatory agents acting by inhibiting the mediators of acute inflammation and is believed to be biphasic (Winter *et al* 1962). The first phase is due to the release of histamine, serotonin and kinin in the first hour after the administration of carageenan; a more pronounced second phase is attributed to the release of bradykinin, protease, prostaglandin and lysosome (Katzung *et al.*, 1998). The later phase of edema is recorded to be sensitive to most of the clinically effective anti-inflammatory agents (Smuker, 1967). The inflammatory granuloma is a typical feature of established chronic inflammatory reaction (Spector, 1969). Turpentine oil-induced granuloma pouch offers a model for exudative type of inflammation (Banerjee, 2000). Ismail et al (1997) found that 1000 mg/kg of root bark powder of *Salacia oblonga* and leaf powder of *Azima tetracantha* to be anti-inflammatory by reducing paw edema volume in the acute phase and reducing the granuloma and exudates in the chronic phase of rats. Though there are numerous studies on the anti-inflammatory therapeutic property of extracts from variety of plants, very few studies have been made on the sources from animal origin. Yegnanarayan *et al.*, (1988) found earthworms to have anti-inflammatory properties and the maximum anti-inflammatory activity in 160 mg/kg of total EP extracted from petroleum ether than from other solvents like benzene, chloroform and ether. The petroleum ether extract significantly reduced the paw edema volume in the acute phase and significantly reduced granuloma pouch weight on cotton pellet induced chronic phase inflammation in rats. Ismail *et al* (1988) also found petroleum ether fraction of total EPs of *Lampito mauritii* to have better anti-inflammatory properties on albino rat and they found 160 mg/kg total EP to function similar to that of aspirin in carageenan induced edema. In our present study 80 mg/kg was found to be more effective. Though EP has been shown to have anti-inflammatory property, the most potent species of worm, dose and the mechanism of action are not clearly understood and needs further investigations.

Free radical scavenging activity was measured by the DPPH scavenging method. Free radicals cause oxidative damage to the body. Antioxidants play a significant role in the body's defence system against free radicals. Recently antioxidants and compounds with radical scavenging activity were reported to be present in fruits, vegetable, herbs and cereal extracts (Gray *at al.*, 2002; Nuutila *et al.*, 2003). DPPH, a stable free radical accepts an electron or hydrogen to become a stable diamagnetic molecule. Present study shows earthworm paste of *L. mauritii* to have scavenging capacities. The data indicates that earthworm paste has a noticeable effect on scavenging free radicals. The effect of antioxidants on DPPH radical scavenging in thought to be due to the hydrogen donating ability.

References

Balamurugan, M., Parthasarathi, K., Cooper, E.L. And Ranganathan, L.S. 2007: Earthworm paste (*Lampito mauritii*, Kinberg) alters inflammatory, oxidative, haematological and serum biochemical indices of inflamed rat. Eur. Rev. Med. Pharmacol. Sci. 11(1): 77-90.

Banerjee, S., Tapaskumar, S., Mondal, S., Chandradas, P. and Sikdar, S.2000: Assessment of the anti-inflammatory effects of *Swertia chirata* in acute and chronic experimental models in male albino rats. Indian J. Pharmacol. 32: 21-24.

Bockus, H.L.1963: Gastroenterology. W.B. Saunder's Company, Philadelphia, USA, p 582.

Bose, M., Mofghare, V.M., Dalchale, G.N. and Turankar, A.V. 2003: Antiulcer activity of Levcromakalim and Nicorandil in albino rats: A comparative study. Pol. J. Pharmacol. 55: 91-95.

Davenport, H.W.1969: Gastric mucosal haemorrhage in dogs–Effect of acid, aspirin and alcohol. Gastroenterology. 56: 439-449.

Duh, P.D. 1998: Antioxidant activity of budrock (*Arctium lappa* L.): Its scavenging effect on free radical and active oxygen. J. American Oil Chemist's Society 75(4): 455–61.

Ebadi, M.2002: Pharmacodynamic basis of herbal medicine. In: Flavonoids. CRC Press, New York.

Costa-Neto Eraldo M.2005: Animal based medicines: biological prospect ion and the sustainable use of Zoo therapeutic resources. *Anais da Academia Brasileira de Ciencias*. 77(1): 33-43.

Goel, R.K., Gupta, S., Shankar, R. and Sanyal, A.K. 1986: Antiulcerogenic effects of banana powder (*Musa sapientum* var paradisiacal) and its effect on mucosal resistance. J. Ethnopharmacol. 18: 33-44.

Gray, D.A., Clarke, M.J., Baux, C., Bunting, J.P. and Salter, A.M. 2002: Antioxidant activity of oat extracts added to human LDL particles and in free radical trapping assays. J. Cereal Sci. 36: 209–18.

Ismail, S.A., Pulandiran, K. and Yegnanarayan, R. 1992:Anti-inflammatory activity of earthworm extracts. Soil Biol. Biochem. 24: 1253-1254.

Ismail, T.S., Gopalakrishnan, S., Hazeena Begum V. and Elango, V. 1992:Antiinflammatory activity of *Salacia oblonga* wall and *Azima tetracantha* Lam. J. Ethnopharmacol. 56: 145-152.

Katzung, B.G. 1998: Basic and Clinical Pharmacology. 7th ed. Stanford: Connecticut. p.578-579.

Martin, M.J., Alarcon, D.E., La Lastra, C., Mashuenda, E., Delgado, F. and Torreblanca, J. 1988: Antiulcerogenicity of the flavonoids fraction from Dittrichia viscose (L). W. Greuter. Phytotherapy Res .2: 183-186.

Middleton, E. Jr. 1996: Biological properties of flavonoids: An overview. Intl J Pharmacognosy 34: 344-348.

Mitsuda, H., Yuasumoto, K., & Iwami, K. 1996: Antioxidation action of indole compounds during the autoxidation of linoleic acid. Eiyo to Shokuryo. 19: 210–214.

Nuutila, A.M., Pimia, R.P., Aarni, M. and Caldenty, KMO 2003: Comparison of antioxidant activities of onion and garlic extracts by inhibition of lipid per oxidation and radical scavenging activity. Food Chemistry. 81: 485–93.

Oboh, G. 2005: Effect of blanching on the antioxidant properties of some tropical green leafy vegetables. LWT, 38, 513–517.

Parmar, N.S., Hennings, G. and Gulati, O.P. 1984: The Gastric antisecretory activity of 3-methoxy 5, 7, 3, 4-tetra hydroxyl flavan (ME)–A specific histidine decarboxylase inhibitior in rats. Agents Actions .15: 143-145.

Ranganathan, L.S. 2006: Vermibiotechnology–From Soil Health to Human Health. Agrobios, Jodhpur, India.

Rao, S.P., Sathiamoorthy, A. and Sathiamoorthy, S.S. 1995: Effect of angiotensin converting enzyme inhibitor captoprilon gastric ulcer production in pylorus ligated rats. Indian J. Physiol. Pharmacol. 36: 296-298.

Smucker, E., Arrhenius, E. and Hulton, T. 1967:Alternation in microsomal electron transport, oxidative Ndemethylation and azo-dye cleavage in CCl_4 and dimethyl nitrosamine induced liver injury. Biochem .J. 103: 55-64.

Spector, W.G. 1969: The granulomatus inflammatory exudates. International Rev. Exp.Pathol. 8: 1-55.

Stephenson, J. 1930: In Oligochaeta. Claredon Press Oxford, p. 658.

Tanaka, M., Kuei., C.W., Nagashima, Y.and Taguchi, T. 1988: Application of antioxidative Maillard reaction products from histidine and glucose to sardine products. Nippon Suisan Gakkaishi 54(8): 1409–14.

Winter, C.A., Risely., E.A. and Nuss, G.W. 1962:Carageenan induced edema in hind paw of the rats as an assay for anti-inflammatory drugs. Proc. Soc. Exp. Biol .Med. 111: 544-547.

Yegnanarayan, R., Ismail, S.A. and Shortri, D.S. 1988:Anti-inflammatory activity of two earthworm potions in Carageenan pedal edema test in rats. Ind .J. Physiol. Pharmac. 32: 72-74.

Chapter 7

Earthworm Culture - An Ecofriendly Way of Improving Soil Productivity

B.K.Dikshit
Regional Sericultural Research Station, Central Silk Board, P.B.No.9, Koraput-764 020, Orissa.

Earthworms are well known from ancient time for their beneficial activities. They play series of beneficial roles in the soil ecosystem and agriculture viz. turning of soil, burrowing and aerating soil, feeding on agricultural and animal waste and releasing castings into soil etc. Their presence and enough population density indicate the healthy (fertile) status of the soil leading to better growth and productivity of plants. Earthworms belong to class Oligochaeta of Phylum Annelida of Animal Kingdom. They are being superfluous hardy creatures, voracious eaters, weigh about one gram each, live up to 6 months without food and water, bisexual (hermaphrodite) in nature but two individuals are normally required for sexual reproduction, discharges 6 to 8 capsules of eggs by mating every 45 days, their eggs stay without hatching for long time if situation is not conducive, give rise to almost 1500 offspring in a year by a single healthy worm, younger worms multiply after 3 months, do not have eyes or ears, understand movement of predators with their most sensitive nervous system, do not have teeth to grind, swallow the food and grind with few stone particles in the gizzard, discharge castings 6 to 7 times a day almost equal to their body weight, body contract and expand through muscular movement, move freely in dark, thrives well in 20-30 ^{0}C temperature and 40-50 % moisture and self sufficient from birth itself.

There are over 4000 earthworm species of which 500 species (Agrawal, 2005) belonging to about 10 families and 58 genera are reported from India (Reddy, 2005). Earthworms are differentiated on the basis of their morphological characteristic with functional significance and are of three types viz. a) Epigeic, b) Endogeic, and c) Anecic (Bouche, 1977).

a) Epigeic species: Live and flourish in top soil rich in decaying organic waste, consume large quantities of litter, usually have small body size, uniform red or brown pigmentation on body, able to survive in adverse conditions, with short life cycle, produce moderate to high rate of cocoons, capable to build up high population density and are highly suitable for recycling of organic waste.

b) Endogeic species: Live below the top soil in the mineral soil layers, characterized by relatively large sized adults, dark pigmented dorsally and interiorly, consume large quantity of soil mixed with decaying organic matter and humus; build horizontal, continuous and ramifying burrows; tolerant to some climatic disturbances, small to large in body size, with intermediate life cycle, produce high rate of cocoons; and play important role in soil formation process, root decomposition, soil mixing and aeration.

c) Anecic species: Deep burrowing species, construct permanent vertical burrows, emerge at night (night crawlers) to feed on organic debris and cast at surface, lightly pigmented at anterior and posterior ends, intolerant to climatic upsets, with long life cycle, produce lower rate of cocoons; and useful for organic matter decomposition, nutrient cycling and soil formation.

Both burrowing (Endogeic and Anecic) and non-burrowing (Epigeic) types of earthworms play significant role in organic farming and sustainable agriculture. Conservation and culture of burrowing worms under *in situ* (in soil) conditions is called as vermi-conservation technology, while mass and *ex situ* (off-soil) culture of epigeic earthworms is called vermicomposting technology.

Earthworms and their feed

The earthworms feed selectively showing preference to a wide variety of decaying plant and organic waste materials, small organisms living or dead and micro organisms present in soil or organic matter. Their sense of taste was reported long back (Darwin, 1881). They always choose wet materials over dried waste, and are influenced by certain physical and chemical properties of the waste material (Reddy, 2005). The palatability of waste materials by earthworms is positively correlated with the nitrogen and soluble carbohydrate contents and negatively correlated with polyphenol like tannin (Lee, 1985). They have a well-organised digestive system. The organic matter when consumed undergoes biological, physical and chemical changes while passing through their gut. The enzymes secreted at different regions of gut help in the digestion of carbohydrates, proteins and fats. The gut of an earthworm acts as a bioreactor, providing ideal conditions of temperature, pH and oxygen concentration for speedy growth of useful aerobic bacteria and actinomycetes resulting in the microbial density. Several environmental factors like oxygen, moisture, temperature, nutrients, and pH influence the microbial population and there by the decomposition process (Ghosh *et al.*, 2005).

Earthworms and vermicomposting

Vermicomposting (derived from Latin word "vermes" means worm) is the process of recycling organic matter into nutrient rich organic compost using earthworms. In vermicomposting, earthworms feed on decomposing organic matter to excrete digested and partially digested matter, as vermicast those put together are known as vermicompost. The renowned naturalist Charles Darwin conducted first scientific observations on earthworms over four decades and published the observations (Darwin, 1881). The first International Symposium on Earthworm Ecology was held at Grange-over-sands in Cumbria, United Kingdom in 1981. In India, vermicomposting technology is about two and half decades old. Vermicomposting enables to link the urban and rural areas from two different angles to reach at a point of developmental process. As the earthworms prefer mixtures of feed rather than a single type of residue, vermicomposting can be very well utilized to decompose mixtures of agricultural, sericultural and urban or industrial organic wastes (Jones *et al.*, 2004). Vermicomposting technology has been accepted by small farmers, large-scale commercial units, agro-based industries, Government Departments and Municipal Corporations for the management of various types of organic materials. Many non-Government institutions and environmental bodies are promoting the activity as an eco-friendly programme to abate organic pollution and to improve soil productivity. Vermicomposting is simple organism (earthworm)-based, cost-effective, affordable and user-friendly biotechnology. Only a few non-diapausing species of epigeic group are used in vermicomposting in India viz.

1. ***Eudrilus eugeniae*** (Kinberg), is tropical but exotic species, originated from Gulf of Guinea in West Africa, large prolific African worm, known as African night crawler; has higher feeding, growth,

reproduction and biodegradation capacity compared to other species; considered as the best among others; more suitable under tropical / subtropical regions; cannot tolerate extended periods below 16 ^{0}C; and suitable for vermicomposting of agricultural crop residues. The occurrence of *E. eugeniae* (Kinberg) was reported from India as early as 1923 but its importance in conversion of organic matter to compost became a new found fact.

2. *Eisenia fetida* (Savigny), is temperate but exotic species, originated from Caucasus region, with red-yellow body colour on dorsal side, known as tiger worm and red wriggler, size of small to medium range with adults of 10 cm long, hatch several young ones per cocoon, tough, survive in mixed species cultures, commonly used in vermiculture, active in wide range of temperature and moisture conditions, and suitable for vermicomposting urban wastes.

3. *Perionyx excavatus* (Perrier), is a tropical species, distributed in India and South-East Asia, known as Indian blue, discovered in a biogas slurry pit in a farmyard garden in the year 1978 in and around Bangalore, India.

4. *Perionyx sansibaricus*
5. *Dendrobaena venata*
6. *Polypheretima clongate(Erseus)*
7. *Octochaetona phillotti* (Mich.)
8. *Octnochaeta rosea* (Stephenson)
9. *Lumbricus terrestris*
10. *Lumbricus rubellus*

However, one endogeic species *Lampito mauritii* (Kinberg), is used for vermicomposting in ground pits in Tamil Nadu, India. In south India with a tropical climate mostly use four-earthworm species viz. *E. eugeniae* (Kinberg), *Eisenia fetida* (Savigny), *Perionyx excavatus* (Perrier) and *Lampito mauritii* (Kinberg), while northern part of the country with a temperate climate *Eisenia fetida* (Savigny), is used for vermicomposting. The earthworm species useful for vermicomposting possess certain inherent biological and ecological characteristics:

- Ability to colonize upper plant-litter layer and organic waste material naturally.
- Showing preference and tolerance to decayed plant materials rich in protein.

- Have high rate of consumption, digestion and assimilation of organic matter.
- Mesothermic nature, using warmth for keeping up of a rapid metabolism.
- Have high reproductive and growth rate producing large number of cocoons.
- Showing high adaptability to wide range of environmental conditions.

Earthworms and soil biology

Earthworms are actively involved in recycling and composting organic waste of the soil either as such or mixed with soil and release vermicast rich in micronutrients to the soil. They alter physical, chemical and biological properties of soil and favour growth and productivity of plants. Earthworms of acidic soil have two calcified glands, which add calcium into the feed. Earthworms along with other cellulolytic and lignolytic micro organisms play a great role in vermicomposting as major decomposers and are boon in the agriculture sector. The vermicast they produce are rich in nutrients, with a low C: N ratio, high porosity and water holding capacity, readily assimilable macro and micro-nutrients, high levels of plant growth regulating substances like Cytokinins and Auxins in amounts comparable with those recorded in soil and rhizospheres, rich Vesicular Arbuscular Mycorrhizal Propagules which play a vital role in the development of plants (Kale, 2005), and higher amounts of humic substances that have hormone like effects on plants (Ghosh *et. al.*, 2005). The cast released by the earthworms into soil activates local earthworm population growth thereby the earthworms burrow through the soil apart by breaking down the root mat and open up the channels for oxygen which harvest water for the soil during rainfall. It is estimated that an earthworm can carve 10 meter long tunnels every day and around 2 Lakh km of tunnels in an acre land in one year (Reddy, N. 2005). The tunnels created by the earthworms are coated with earthworm mucus rich in nitrates are available to the soil. Plant roots can extend quickly along these channels resulting in healthy growth. The plants' residue mostly has a carbon-nitrogen (C: N) ratio of greater than 20:1. At this level nitrogen cannot be absorbed by plants, which makes soil acidic. To bring down this ratio, earthworms play important role by consuming the biomass and converting it into valuable manure (Pai, 2005). In soil earthworms work in collaboration with aerobic bacteria and other organisms to convert organic waste into compost and prevent the growth and multiplication of harmful anaerobic bacteria, other pathogenic organisms and vectors of diseases by creating oxygen rich (aerobic) environment. Further, antibiotics viz. Streptomycin, Terramycin, and Erythromycin are found in the environment of earthworms. Earthworms releasing their cast into soil showed bacterial count in soil 1000 times more than without vermicompost (Pai, 2005), steady decline in termite population, and free from weed seeds as the food is churned out to fine particles like "micro mills" with their gizzards present in intestine before passing as excreta. It is also reported that earthworms play a significant role to recharge ground water and prevent runoff losses. The vermiwash a by-product of vermicomposting is a coelomic fluid of earthworms and can be used as spray to nursery or field crops & in the tissue culture medium as mineral and vitamin substitute and its quality can be maintained at room temperature by using some organic extracts and other active ingredients (Kale and Radhakrishna, 2005). Amount of nutrition multiplied by earthworms from the matter they feed is Nitrogen by 5 times, Phosphorous by 7 times, Potash by 11 times, Magnesium by 2.5 times and Calcium by 2 times (Reddy, N.2005). After death also the earthworms decay and add fertility to soil.

Conclusion

Soil, the biologically active and porous medium, is the principal substrata of life on earth and serves as a reservoir of water and nutrients. Plants receive all macro and micro nutrients from soil. Modern agriculture largely relies on artificial means with liberal use of chemical fertilizers and pesticides, hazardous effects of which are now being realized. Large amounts of toxic residues and metabolites are released in the environment that resulting contamination (pollution) of natural resources including soil, air and water, depletion of population of beneficial flora and fauna of soil, decline in soil fertility, uncertainty and stagnation of crop productivity, contamination of food products and deterioration of their quality, health hazards and new genetic disorders. In addition to all these, bovine wealth of farmers is fast declining thereby affecting the production of farmyard / cow dung manure for field use. Besides, cow dung is used as fuel by most of the farmers. As such, recycling of agricultural farm wastes by vermiculture technology should receive greater attention. Earthworms are small creatures, which do their work silently and out of sight. They were used all these years for fishing and eating. Although they render excellent service, we have to look beyond this usage. They are best in restructuring the soil, abundantly available at no cost. For better exploitation of the potential of earthworms to produce maximum vermicast, earthworm farming is imperative.

References

Agrawal, O.P. 2005: Vermitechnology for all round sustainable Development. National Seminar on Composting and Vermicomposting (Paper presented at), RSRS, Central Silk Board, Kodathi, Bangalore, p. 39-47.

Bouche, M.B. 1977: Strategies of lombriciennes. *Biol. Bull.* 25: 122-132.

Darwin, C. 1881: The formation of vegetable mould through the action of worms with observations of their habits. Murray, London.

Ghosh, B.C., Palit, S. and Banerjee, H. 2005: Development of production and processing technology of vermicompost unit at commercial scale. . National Seminar on Composting and Vermicomposting (Papers presented at), RSRS, Central Silk Board, Kodathi, Bangalore, p. 36-38.

Giraddi, R.S. 2005: Structural designs for production of Vermicompost using agricultural and other farm wastes. National Seminar on Composting and Vermicomposting. (Papers presented at), RSRS, Central Silk Board, Kodathi, Bangalore. p. 30-35.

Jones Nirmalnath, P., Biradar, A.P., Patil, A. B. and Patil, M.B. 2004: Vermicompost micro flora as influenced by different crop residues. Karnataka .*J. Agri. Sci.* 19(1): 168-169.

Kale, R.D. 1995: Vermicomposting has a bright scope. *Indian Silk.* September. Bangalore. p.6-9.

Kale, R.D. and Radhakrishna, D. 2005: Development of Vermicomposting Technologies in India- Present position and future prospective. National Seminar on Composting and Vermicomposting. RSRS, Central Silk Board, Kodathi, Bangalore, p. 17-21.

Lee, K.E. 1985: Earthworm: their ecological relationship with soil and land use. Academic Press, New York.

Pai, S. 2005: Earthworm farming- an ecofriendly approach for sustainable soil fertility management and crop productivity. National Seminar on Composting and Vermicomposting (Papers presented at), RSRS, Central Silk Board, Kodathi, Bangalore. p. 65-70.

Reddy, M.V. 2005: Vermicomposting using different species of earthworms - A comparative account. National Seminar on Composting and Vermicomposting, RSRS, Central Silk Board, Kodathi, Bangalore, p. 22-25.

Reddy, N.L. 2005: Sharing my experience in vermiculture. National Seminar on Composting and Vermicomposting, RSRS, Central Silk Board, Kodathi, Bangalore. p. 79-81.

PART - III

EARTHWORMS IN AGRO ECOSYSTEMS

Chapter **8**

Impact of Rubber (*Hevea brasiliensis*) Plantation on the Earthworm Communities in Tripura (India)

P.S.Chaudhuri and Subhalaxmi Bhattacharjee
Department of Zoology, Maharaja Bir Bikram College, Agartala-799 004 (India)

In tropical and sub-tropical areas, tree plantations are becoming an increasingly common land use system. Plantations are being established for different reasons including shift in timber production from native forests to plantations, restoration of degraded lands, catalysts of forest succession and also as buffer zones for biodiversity conservation (Tien *et. al.*, 2000; Sarlo, 2006).

In Tripura, a north eastern state of India, rubber (*Hevea brasiliensis*) plantation was introduced in 1963 by the Forest Department to check soil degradation due to slash and burn practiced by the local tribal people and also as a part of their rehabilitation programme. Tripura, having a sub-tropical climate, occupies the second position in area and productivity among the rubber growing states of India with 33,000 hectare of land under rubber cultivation (Bahuguna, 2006). *Hevea brasiliensis* (Rubber) is an exotic plant with its origin in Brazil. Being a deciduous plant with very fast rate of growth, it shows maximum litter fall during February and March with annual litter addition to plantation floor amounting to 7 ton per hectare (Jacob, 2000). The litters are not generally removed but persist on the plantation floor throughout the year and show very slow rate of decomposition due to high lignin content. Some phenolic compounds are known to be present in the rubber plant material (Stern, 1967).

Direct effect of plant species on soil and litter biota are caused by the plant's inputs of organic matter above and below the ground; while indirect effect of plants on biota includes shading, soil protection and uptake of water and nutrients by the roots (Neher, 1999). That individual tree species rather than monoculture or polyculture had a significant effect on earthworm biomass was reported by Sarlo (2006). González *et al.*, 1996 and Tien *et al.*, 2000, reported that tree plantations may influence earthworm abundance by altering the physico-chemical properties of soils viz. temperature, moisture regime, pH, soil organic matter content and litter inputs. Moreover, native tree plantations generally allow the establishment of populations of native earthworm species (Zou and González, 2002). Several authors suggest that the chemical composition of plants, especially N, lignin contents and phenolic compounds play a vital role in the abundance of soil and litter fauna through their effect on palatability and rate of decomposition (Tien *et al.*, 2000 and Sarlo, 2006). The quality and quantity of food material influences not only the size but also the species composition, growth rate, fecundity of an earthworm population (Chaudhuri *et al.*, 2003). They have reported differences in the rate of growth and reproduction of three vermicomposting species, *Perionyx excavatus, Eudrilus eugeniae* and *Eisenia fetida* in the *Hevea* leaf litters used as vermiculture substrate. Earthworms are known to affect the soil physico-chemical and biological properties to a large extent. Moreover, they constitute the highest biomass among the tropical soil macrofauna. Therefore, their role in ecosystem function, biodiversity conservation and restoration of degraded soil is of great importance.

Although there are reports on the occurrence of microarthropods, especially oribatid mites in the *Hevea* soils of Tripura (Bhattacharjee and Chakraborty, 1995) but probably no records on distribution and diversity of earthworms in the rubber growing areas of India including north-eastern states that constitute a huge reservoir of biodiversity.

This present paper deals with the impact of rubber plantation on earthworm communities (density, biomass etc), relationship between density and biomass of earthworm community with edaphic factors (moisture, temperature etc.) of the rubber plantation and also the distribution of earthworm species within the plantation with reference to the edaphic factors.

Study site

The study was carried out at Taranagar Rubber Research Farm (23° 53' N and 91° 15' E) of Mohanpur Block, West Tripura and falls under sub-tropical deciduous forest region with annual rainfall ranging from 1500-2000 mm. This is a considerably large area (86 ha) comprising of rubber plantations of different age groups. For the purpose of our study, three *Hevea* plantations of different age groups (3yr, 10 yr and 25yr old) were selected. Both the mature plantations (10yr and 25 yr) possess canopy cover; horizontal distribution of roots and the plantation floors remain covered with *Hevea* leaf litter. The juvenile plantation, however, has almost no horizontal root distribution and no canopy cover. The floor of the juvenile plantation shows the presence of leguminous cover crop, grasses etc with scanty distribution of *Hevea* leaf litter. Unlike the mature plantations, the juvenile one shows a dramatic rise in ant (both red and white ants) population. The climatic features of the region include summer (April-June), monsoon (July-October), winter (November-February) and a short spring (March) with an annual rainfall of about 1814.50mm. The soil is Ultisol, well drained, highly acidic (pH 4.3-5.3) with sandy loam/ sandy clay loam/ clay loam texture. Mineral (clay) content of Taranagar Rubber Research Farm are Kaolin (56%), Illite (10%), vermiculite (5%), Interstratified (5%), alkali fildspass (2%), silica (21%) and Goethite (1%). Highly acidic nature of soil is indication of the fact that these soils have developed from non-calcareous parent material under conditions of high rainfall (Chaudhuri *et al.*, 2004).

Sampling of earthworm population

Sampling was done every month at *Hevea* plantations of different age groups (3 yr, 10 yr and 25 yr) for a period of seven months (June to December, 2006). From each of the three plantations under study, an area of 1 ha was chosen within which 5 widely separated sampling plots (10m x 10m) were randomly selected every month. Composite sample, comprising of five sub samples, was taken from each of them. The sub samples were located at the 4 corners and the center of each sampling plot. Earthworms were collected by the conventional digging and handsorting method. Worms were counted, weighed and some were preserved in 4% formalin. Preserved specimens were sent to The Zoological Survey of India, Solan (Himachal Pradesh) for identification. For conservation of biodiversity, minimum number of worms (10-15) was preserved and others were released back to the soil. Results were expressed in terms of biomass [fresh wt (g)] and density (individuals/m^2)

Soil analysis

Soil temperature (soil thermometer), soil moisture (gravimetric method) were recorded at each sampling site from depth of 0-15cm. Soil pH (1:5 water dilution method, Electrometric method), organic carbon (Walkley and Black rapid titration method), Potassium and Calcium (Flame Photometric method), nitrogen (Micro-Kjeldahl method), Phosphorus (Troug's method, UV-VIS Spectrophotometer) and water holding capacity (Upadhayay and Sharma, 2001) were analysed.

Vegetation analysis

Various shrubs, herbs etc. which were found on the plantation floors, were collected and identified. To estimate the difference in amount of leaf litter, average dry weight of leaves (g/ m^2) from 10 samples was taken from each of the three plantation floors during the month of March.

Casts collection

Casts were collected only during peak of earthworm activity i.e. September-October. Average dry weight (g/ m^2) of casts collected from 10 samples was recorded from each of the three plantations.

Statistical analysis

t-test was used to compare significant difference in density (number of individuals/m^2) and biomass (g/m^2) of earthworms among different age group of plantation. The correlation between the ecological factors (moisture and temperature) and density, ecological factors and biomass of earthworm in 25 yr old *Hevea* plantation were graphically represented showing scattered diagram and the best fitted straight line.

Experimental findings

Earthworms were concentrated mainly in the first 0-15 cm of soil depth. Peak of earthworm activity (indicated by increase in both biomass and density and also their castings activity) was found during September - October i.e. post-monsoon boom (Figs. 1 & 2). Present study revealed that the total number of earthworms (including juveniles) collected from three sites was 274 (3 yr old plantation), 582 (10 yr old plantation) and 646 (25 yr old plantation) with occurrence of 6 species of earthworms belonging to 5 genera and 4 families in the rubber plantations under study (Table 1). The number of

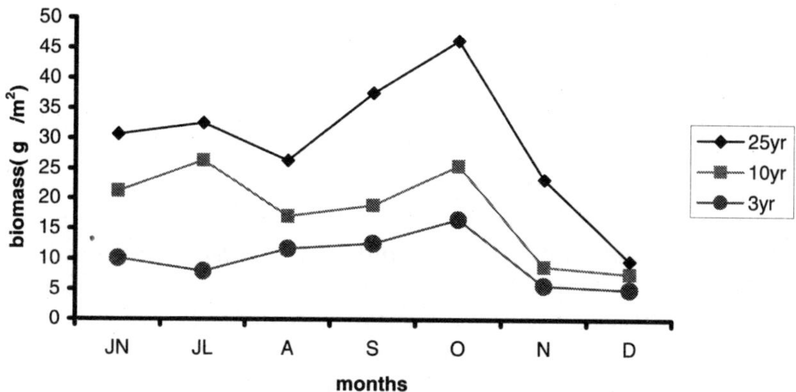

Fig 1: - Monthly fluctuations in average biomass of earthworms in 3yr, 10yr, 25 yr old rubber plantation

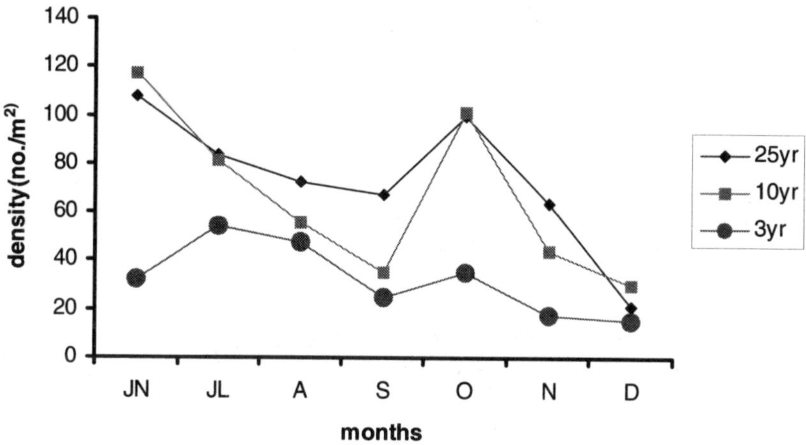

Fig 2:- Monthly fluctuations in average density of earthworms in 3yr, 10yr, 25 yr old rubber plantation

earthworm species is expected to increase after the completion of identification. Among the 6 species 2 were from the family Octochaetidae (*Eutyphoeus comillahnus, Dichogaster* sp), 2 from the Moniligastridae (*Drawida nepalensis, Drawida* spA), 1 each from the Megascolecidae (*Metaphire houlleti*) and the Glossoscolecidae (*Pontoscolex corethrurus*).

Peregrine species like *Pontoscolex corethrurus, Drawida* sp, *Metaphire houlleti* showed abundance and were found in all of the three plantations under study; while the adults of *P. corethrurus* were found in the *Hevea* plantations throughout the year, those of *M. houlleti* showed restricted period of activity with sudden disappearance in the month of November. *Pontoscolex* sp. is well known for its wide range of moisture tolerance and capacity of more resistance to dehydration in the soils (Edwards and Bohlen, 1996). These are probably the main reasons for wide range of activity period of this species in the soil of rubber plantations. *M. houlleti* is a litter feeder. So, its activity period must be related to the stages of decomposition of the *Hevea* leaf litters. Interestingly, *M. houlleti* was found to be distributed in huge number in the over ground heaps of decomposed *Hevea* leaf litters (unpublished data). Native worm, *Eutyphoeus comillahnus* was found to be restricted within the 25 yr old plantation only. Occurrence of *Eutyphoeus* sp. under climax forest was reported by Bhaduria and Saxena (2007). Kalisz and Dotson (1989) reported that in the deciduous forests in Kentucky, only native worms occurred in the undisturbed or slightly disturbed sites, whereas exotic species occurred in severely disturbed sites. Rubber plantation often faces anthropogenic interferences for fertilizer application, tapping procedure, forest clearing etc. So abundance of exotic species like *P. corethrurus* in the rubber plantation areas is not surprising.

A number of ecological factors viz. soil moisture, pH, organic matter etc. are known to play vital role in the distribution, diversity and abundance of earthworms. Distribution of earthworm species in the three age groups of rubber plantations in relation to different ecological parameters is given in Table 1.

Table 1. Earthworm species in the three rubber plantations (3 yr, 10yr, 25yr) of Taranagar and their ecological conditions

FAMILY AND EARTHWORM SPECIES	Soil Temp (°c)	Soil Moisture (g%)	Soil pH	Soil Organic Matter (g%)
Glossoscolecidae				
1. *Pontoscolex corethrurus*	19.00- 32.20	10.00-29.08	4.54-4.87	1.72-2.42
Megascolecidae				
2. *Metaphire houlleti*	21.00-28.00	13.70-20.35	4.43-5.21	1.64-1.92
Moniligastridae				
3. *Drawida nepalensis*	21.00-30.20	10.00-29.08	4.43-5.21	1.64-2.42
4. *Drawida* sp A	22.00-32.20	10.08-29.08	4.54-4.76	1.52-1.92
Octochaetidae				
5. *Eutyphoeus comillahnus*	21.00-26.00	17.00-18.21	4.58-4.60	1.72-1.84
6. *Dichogaster* sp.	ND	ND	ND	ND

The present investigation revealed that earthworm species of the three plantations generally occurred in moist (10-29%) moisture, highly acidic soil (pH 4.4-5.2) with soil temperature ranging from 19-32°c and organic matter content >1% (Table I). Variation in soil moisture, temperature, pH, nitrogen, potassium, calcium, phosphate levels in the three age groups of plantation is given in Table II b. Comparatively high nitrogen level in 3 yr old plantation was probably due to its ground cover of leguminous plants that fix atmospheric nitrogen. Differences in the nutrient content in *Hevea* soil of different age group plantations might be related to differences in the rate of fertilizer input. According to Chaudhury *et al.* (2001) most of the rubber growing sites of India has low level of available phosphorus, low to medium level of organic carbon and available potassium and medium to high level of available magnesium.

Variations in species richness among the three plantations under study are not clear because all the species inhabiting these plantations have not yet been identified. However, there was a clear trend in increase of both average biomass (3 yr old plantation **9.98g/ m^2**, 10 yr old plantation **17.95 g/ m^2**, 25 yr old plantation **29.51 g/m^2**; n = 35) and density (3 yr old plantation **32.46/ m^2**, 10 yr old plantation **66.51 / m^2**, 25 yr old plantation **73.82/ m^2**; n = 35) with increase in the age of plantation (Table II a). Plantation age related increase in the worm biomass and density was statistically significant (Tables III a & III b). Density and biomass value of earthworm in 25 yr *Hevea* plantation is close to those values obtained by Fragoso and Lavelle (1987) in the tropical rain forests of Mexico. Our observation corroborates with the study of Gillot *et al.* (1995), who also reported gradual increase in the density and biomass of earthworms in the rubber plantations of Côte d' Ivoire with the increase in age of plantation. Significant increase in earthworm density associated with the age of *Eucalyptus* plantation for earthworm density was noticed by Mboukou- Kimbatsa and Bernhard- Reversat (2001). Increase in the amount of leaf litters leading to increase in the amount of organic matter with the increase in age of *Hevea* plantation might have a relationship with increasing earthworm activity in the mature plantation.

Previous studies by Fragoso *et al.* (1997) revealed that trees with relatively fast growth favor the development of earthworms, presumably through their effects on litter and microclimate in the soil. Interestingly, Sarlo (2006), found that biomass of earthworms was significantly correlated with canopy cover. Presence of thick

Table – II a : Biological Characteristics of *Hevea* plantations of different age groups

Age of *Hevea* plantation	**3 yr old**	**10 yr old**	**25 yr old**
Plantation characteristics			
Canopy	Absent	Present	Present
Ground cover	Leguminous plants and grasses	*Hevea* leaf litter	*Hevea* leaf litter
Above ground vegetation	10 species *Melastoma malabaricum, Pureria phaseolides, Arundinella bngalensis, Chromolaena odorata, Triumfetta rhomboidea, Synedrella nodiflora, Hedyotis coronaria, Hyptis suaveolens, Spoermacoce hispida, Mimosa pudica*	6 species *Epimeredi indicus, Desmodium pulchellum, Lygodium flexuosum, Ichnocarpus frutescens, Pureria phaseolids, Clerodandrum viscosum*	6 species *Chromolaena odorata,Mikania cordata, Achyranthus aspera, Curcuma zedoaria, Melastoma malabaricum, Dryopteris* sp
Amount of rubber leaf litter	2.8 g/m^2	0.2-0.7 kg/m^2	0.8 – 1.5 kg/m^2
Earthworm Community Structure			
Number of earthworm species till identified	5	5	6
Name of earthworm species	*Pontoscolex corethrurus, Drawida nepalensis, Drawida sp A, Metaphire houlleti, Dichogaster sp*	*Pontoscolex corethrurus, Drawida nepalensis, Drawida sp A, Metaphire houlleti, Dichogaster sp*	*Pontoscolex corethrurus, Drawida nepalensis, Drawida sp A, Metaphire houlleti, Dichogaster* sp, *Eutyphoeus comillahnus*
Avg dry wt of cast (g/m^2)	2.0	67.0	83.7
Total no. of worm collected (N=35)	274	582	646
Avg. biomass (g/m^2)	9.98	17.95	29.51
Avg. density (no. / m^2)	32.46	66.51	73.82

Table - II b : *Edaphic Characteristics of Hevea plantations of different age groups*

Edaphic characters	3 yr old	10 yr old	25 yr old
(n= 5)			
Soil moisture (%)	15.29	15.31	17.08
Soil temperature (°c)	26.22	24.45	24.39
pH	4.88	4.58	4.59
Organic matter (g%)	1.75	2.06	1.80
Nitrogen (ppm)	91.60	76.20	84.00
Potassium (mg/100g)	1.476	1.772	1.496
Calcium (mg/100g)	161.176	163.252	105.164
Phosphate (mg/100g)	5.6208	4.5592	4.964
Water holding capacity (%)	43.00	42.91	45.98
Human interference	++	++++	+
Fertilizer application	+	+++	-

canopy helps to cut down direct radiation, minimize evaporation run off in the 25 yr old plantation. Beside these, reduced soil temperature in this age group of plantation not only improves soil moisture status but also leads to reduced oxidation of soil organic matter and favors its build up. Thus in our present study on 25 yr old *Hevea* plantation, both moisture and temperature had a strong correlation with both the biomass and density of earthworms (Fig. 3a & 3b; 4a & 4b). Tiwari *et al.* (1992), also found a significant correlation between earthworm populations and edaphic factors like temperature and moisture in a pineapple field. According to Edwards and Bohlen (1996), moisture and temperature of soil can act synergistically to influence earthworm activity. Amount of surface casts increased with age of rubber plantation (3yr plantation **2g m^{-2}**, 10yr plantation **67g m^{-2}**, 25yr plantation **84g m^{-2}**) during post monsoon period. Increase in the castings activity with the age of plantation was probably related to increase in shaded areas (provided by the canopy) with improvement in soil moisture status in the mature plantations. Roy (1957) reported annual earthworm cast production ranging from 1.6 to 31.0 tons / acre in the grasslands at Kolkata (India). It is probable that most of the earthworm casts in *Hevea* soil were deposited subsurface due to less compactness of soil caused by extensive horizontal distribution of roots. Significantly low density and biomass value of earthworms and minimum castings activity in 3 yr old plantation compared to mature plantations is probably due to its high temperature and low

25 yr plantation

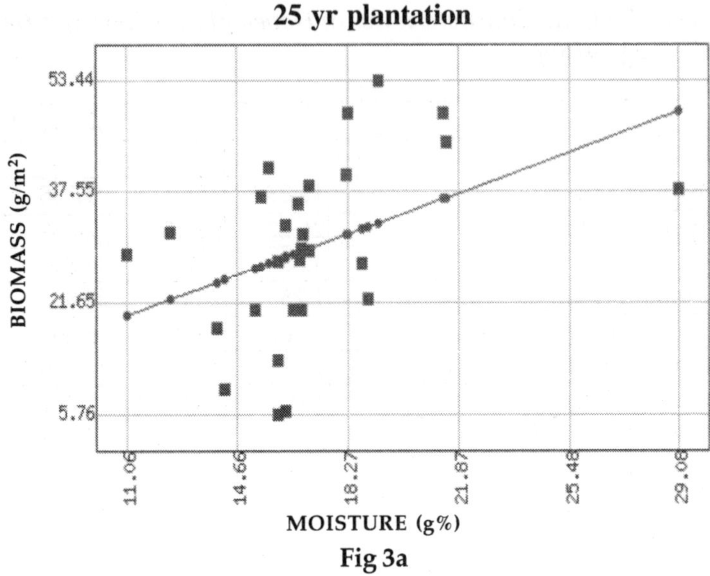

Fig 3a

(Y= 1.7767+1.623X, r = 0.4192, t = 2.3542, p = 0.0264)

25 yr plantation

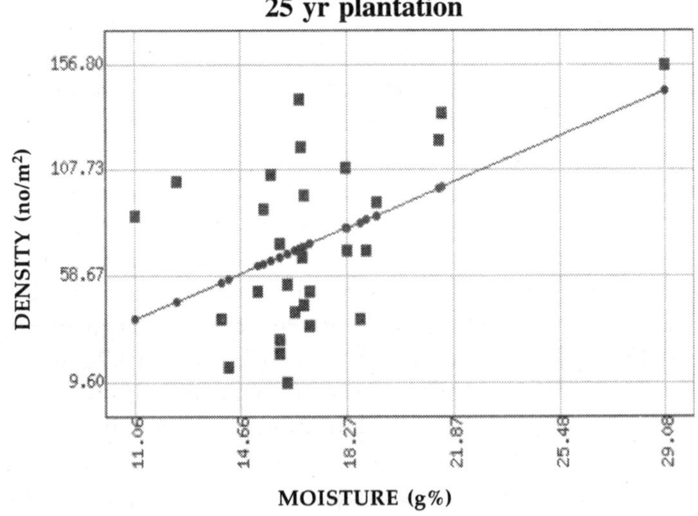

Fig. 3b

(Y= -26.8758+ 5.8948X, r = 0.4783, t = 2.7769, p = 0.010041)

Fig 3 (a & b): - Relationship between density and biomass of earthworms with soil moisture (a) moisture vs. biomass (b) moisture vs. density in the 25 yr old *Hevea* plantation

Table– III a : *Test of mean differences in earthworm biomass among different age groups of* Hevea *plantations*

Parameter	*Plantation age groups*	t value	p value
EARTHWORM BIOMASS	*Between 3yr and 10yr plantation*	5.98071	0.000064
	Between 3yr and 25yr plantation	12.58081	0.00000003
	Between 10yr and 25yr plantation	6.76114	0.00002

Table– III b : *Test of mean differences in earthworm densities among different age groups of* Hevea *plantations*

Parameter	*Plantation age groups*	t value	p value
EARTHWORM DENSITY	*Between 3yr and 10yr plantation*	12.42213	0.00000003
	Between 3yr and 25yr plantation	16.04667	0.000000001
	Between 10yr and 25yr plantation	2.35978	0.0361

moisture content of the soil in absence of canopy cover in spite of the fact that here plantation floor is covered with leguminous plants to provide nitrogen to the soil. Difference in canopy cover (affecting solar radiation), quality (related to concentration of tannin and related polyphenolic compounds, C: N ratio etc.) and quantity of leaf litter, biotic resistance, anthropogenic influence, various ecological factors etc. might have triggered changes in the abundance and community structure of earthworms among the three plantations.

25 yr plantation

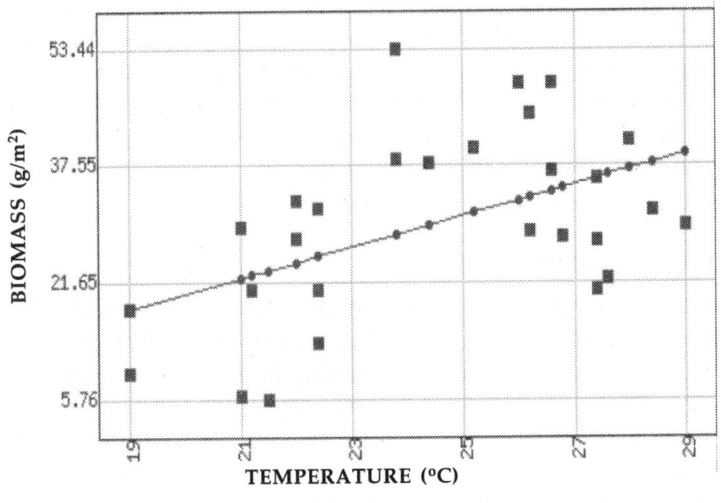

Fig. 4a

(Y= -22.3179+2.1194X, r = 0.4998, t = 2.9427, p = 0.0067)

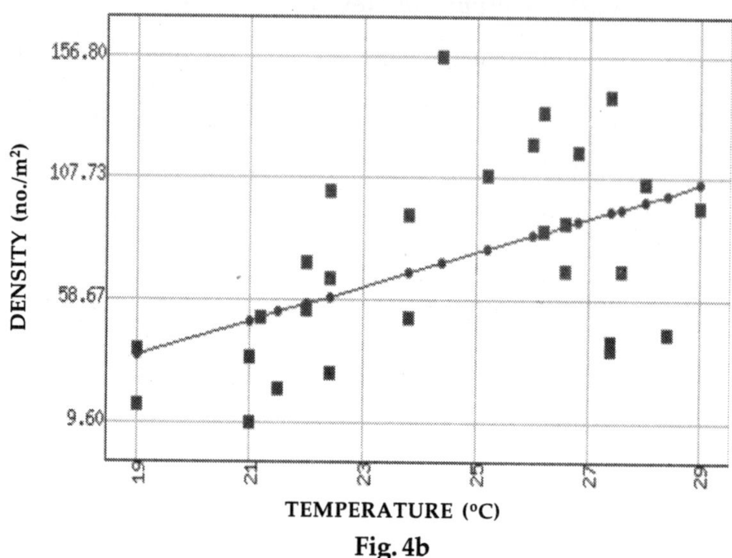

Fig. 4b

(Y= -94.3208 + 6.8763X, r =0.5095, t = 3.0197, p = 0.0056)

Fig 4(a & b): - Relationship between density and biomass of earthworms with soil temperature (a) temperature vs. biomass (b) temperature vs. density in the 25 yr old *Hevea* plantation

In fine, gradual increase in earthworm population with increase in plantation age indicates that *Hevea* plantation supports and improves the earthworm population through changes in the microclimatic conditions of the soil. Further research is on progress to study the structure of earthworm community in the neighbouring unutilized fallow lands. Thus in Tripura, conversion of unutilized fallow lands to rubber plantation will tell about the role of such social forestry in the conservation of soil biodiversity.

Acknowledgement

The authors express their sincere thanks to Dr. R. Paliwal, ZSI, Solan for the identification of earthworm species, Dr. Nalinikanta Chakraborty, M.B.B College, Agartala, Tripura, for identification of different plant species of rubber plantation, Dr. Ratan Kumar Saha, College of Fisheries, Lembucherra, Tripura for chemical analysis of soil and Ratul Chakraborty, M.B.B College for statistical analysis. The authors also thank Department of Science and Technology, New Delhi, for sponsoring the study and Dr. S.K Dey, Rubber Research

Station, Agartala for allowing the survey work at Taranagar Rubber Research Farm, Mohanpur Block, West Tripura.

References

Bahuguna, V.K. 2006 : Natural Rubber in Tripura. Technical Bulletin 1, Tripura Rubber Mission, Govt. of Tripura.

Bhaduria, T. and Saxena, K.G. 2007: Influence of Landscape modification on earthworm biodiversity and community in garhwal region of Central Himalayas. Indo-US Workshop on "Vermitechnology in Human Welfare" (abstract), Coimbatore.

Bhattacharjee, T. and Chakraborty, P. 1995: Community structure of soil Oribatida of a young rubber plantation and an adjacent wasteland in Tripura (India). In: Advances in Ecology and Environmental Sciences (Ed. P.C Mishra, N. Bahera, B.K. Senapati and B.C. Guru), Ashish Publishing House, New Delhi, p. 65-77.

Chaudhury, M., Sarma, A.C., Pal. T.K., Chakraborty, S.K. and Dey, S.K. 2001:Available nutrient status of the rubber (*Hevea brasiliensis*) growing soils of Tripura. Indian Journal of Natural Rubber Research, 14(1): 66-70.

Chaudhury, M., Sarma, A.C., Pal. T.K., Chakraborty, S.K. and Dey, S.K. 2004:Available nutrient status of the rubber growing soils in the lower Brahmaputra valley of Assam. Natural Rubber Res. 17(2): 177-179.

Chaudhuri, P.S., Pal, T.K., Bhattacharjee, G. and Dey, S.K. 2003: Rubber leaf litters (*Hevea brasiliensis*, var RRIM 600) as vermiculture substrate for epigeic earthworms, *Perionyx excavatus, Eudrilus eugeniae* and *Eisenia fetida*. Pedobiologia, 47: 796-800.

Edwards, C.A. and Bohlen, P.J. 1996: Biology and Ecology of earthworms. Chapman and Hall, London.

Fragoso, C. and Lavelle, P.1987: The earthworm community of a tropical rain forest, In: On Earthworms (Eds.A.M Bonvicini- Pagliani & P. Omodeo), Modena Italy, p.281-295.

Fragoso, C., Brown, G., Patrón, J.C., Blanchart, E., Lavelle, P., Pashanasi, B., Senapati, B.K. and Kumar, T. 1997: Agricultural intensification, soil biodiversity and agro ecosystem function in the tropics : the role of earthworms. Applied Soil Ecology 6: 17-35.

Gilot, C., Lavelle, P., Blanchart, E., Keli, J., Kouassi, P. and Guillaume, G. 1995: Biological activity of soil under rubber plantations in Côte d' Ivoire. Acta Zool. Fenica, 196: 186-189.

González, G., Zou, X. M. and Borges, S. 1996: Earthworm abundance and species composition in abandoned tropical croplands: comparisons of tree plantations and secondary forests. Pedobiologia, 40: 385-391.

Jacob, J. 2000: Rubber tree, man and environment. In: Natural Rubber: Agromanagement and crop processing (Eds: P.J George and C.K Jacob), Rubber Board, Kottayam, India, 599-610.

Kalisz, P.J. and Dotson, D.B. 1989: Land use history and the occurrence of exotic earthworms in the mountains of Eastern Kentucky. American Midland Naturalist 122: 288-297.

Mboukou- Kimbatsa I.M.C and Bernhard- Reversat F. 2001: Effect of exotic tree plantations on invertebrate soil macrofauna. In: Effect of Exotic Tree Plantations on plant diversity and biological soil fertility in Congo Savana: with special reference to *Eucalyptus* (Ed: F. Bernhard- Reversat) Center for International Forestry Research, Bogor, Indonesia, p. 49-55.

Neher, D.A. 1999: Soil community composition and ecosystem processs: comparing agricultural ecosystems with natural ecosystems. Agroforestry Systems 45: 159.

Roy, S.K. 1957: Studies on the activities of earthworms. Proc. Zool. Soc. Bengal 10: 81-98.

Sarlo, M.2006: Individual tree species effects on earthworm biomass in a tropical plantation in Panama. Caribbean J. Sci.42 (3): 419-427.

Stern, H.J. 1967: Rubber: Natural and Synthetic. Maclaren and Sons Ltd. London.

Tien, G. Olimah, J.A. Adeoye, G.O. and Kang, B.T.2000: Regeneration of earthworm population in a degraded soil by natural and planted fallows under humid tropical conditions. Soil Sci. Soc. Am. J. 64: 222-228.

Tiwari, S.C., Tiwari, B.K. and Mishra, R.R. 1992: Relationship between seasonal populations of earthworms and abiotic factors in pineapple populations. Proc. Nat. Acad. Sci. Ind. Sec B (Biol. Sci.) 62(2): 223-226.

Upadhayay, R.M.and Sharma, N. L. 2001: Manual of Soil, Plant, Water and Fertilizer Analysis. Kalyani Publishers, New Delhi.

Chapter 9

Economic and Environmental Potential of Tasar Silkworm and Earthworm Interaction in a Forest Ecosystem

K.V.Shankar Rao

Regional Sericultural Research Station, Landiguda. Koraput-764 020 (Orissa), India

The Convention on Biological Diversity (CBD) has identified India as one of the 12 mega biodiversity region very rich in several unique species of flora and fauna and north-eastern India (22°- 29°N and 90°- 97°E) is one of the 18 hot spots known for biodiversity in the world, which is a homeland of several wild silk moths.

Bastar Biosphere Reserve - a stretch of 23,138-sq.km.of tropical dry deciduous, is a reserved forest and is one of the notified biodiversities of India. It is a conglomeration of Kanger National Park, a project tiger area, undulating hills with varied topography, soil, climate, cascading water fills, vast- Sal plant (*Shorea robusta*) forests and myriad's of flora and fauna of diverse genetic resources. Bastar Biosphere Reserve is rich in seri-biodiversities such as unique Sal forest and wild tasar silkworm ' Raily 'producing tough stone like cocoons and helps in income generation for local tribal. The silkworm is polyphagous in nature and distributed between 17° 14' to 20° 34'North latitude and between 80° 15' to 82° 15'East latitude. Raily eco-race development project (1997-2004) of Government of India enhanced cocoon collection to 3843.72 lakh and generated Rs.352.04 crores at different levels and helped the forest dwellers and created awareness for forest conservation .The study also observed the inter-dependence of tasar silkworm and earthworms as litter of the silkworm and plant biomass form feeding material for earthworms, thereby maintaining environmental balance.

Frequency Distribution of Tasar Flora

The studies on plant community structure of tasar flora and their associates in forest of Bastar revealed that the *Shorea robusta* among the tasar flora is a predominant species in the dry sub-humid and moist sub humid zone of forests with a relative frequency of 71.07% and 45.92%, respectively and overall relative frequency of 54.88% in Bastar division. Among its common associates the predominant species is *Terminalia tomentosa* with a relative frequency of 4.04% and 6.54% in the said zones and overall relative frequency of 5.61%. Variation in relative frequency of a particular species in different zones indicates specific adaptive features for its successful survival in the particular environment.

Productivity of Raily Ecorace

There are two seasons for the collection of Raily tasar cocoons. The first seasonal collection starts from August and ends by October and second season from the month of February and are continued till the end of May. In an estimate 300-750 lakhs of Raily cocoons are produced annually in 21 forest ranges of 6 divisions of this central Bastar division accounted for 33.33%, East Bastar 27.45%, North Bastar 21.57%, South Bastar 15.69% and Kanker 1.96%.

Declining Trend of Natural Grown Raily Cocoons

The production trend indicates a faster depletion of the natural population density due to deforestation, human intervention in the forests, changes in the micro environmental conditions and habitat. The high price offered to the wild cocoons has resulted in excessive collection of wild cocoons of Raily and consequently the natural population has declined sharply in its natural habitat.

Conservation

Since Raily wild silkworm is not amenable to human handling, ex-situ conservation in the research centers is not feasible. "Our existing knowledge of this species is not sufficient to handle this species under ex-situ condition. Further continuous breeding and multiplication of wild populations lead to inbreeding depression and subsequent loss of genetic stock and therefore only in-situ conservation is essential.

Different Conservation Systems in-situ

Various natural regression systems suitable to different nature grown eco-pockets were evolved. These includes:

1. Release of seed cocoons.
2. Release of moths.
3. Release of gravid moths.
4. Release of eggs in Sal leaf cups.
5. Release of chawki worms.

All these systems are useful for genetic stock conservation and rejuvenation. Higher profitability index was noticed for the release of eggs in Sal leaf cups (1:2.75) followed by release of chawki worms (1:1.54) and fertilized moths (1:1.34).

Adoption of Conservation Strategy

Based on the preservation, grainage and rearing parameters under in situ, a conservation strategy was formulated. This includes:

1. Identification of potential eco-niches.
2. Augmentation of seeds.
3. Methods of conservation and technology adoption.

Chhattisgarh government adopted in-situ conservation on large scale. A total 29 lakhs of cocoons were utilized and a total of 4.36 lakhs of dfls were released from 1997-2004.

4. Impact on community development

Central government assisted Raily development project and was operated from 1997-2004 which produced 3843 lakhs of cocoons resulted in weaving of 48 lakhs meters of silk fabric. The project overall created 198.35 lakhs of man-days and generated Rs. 352.04 crores of at cocoon, reeling and fabric level. On an average, the strategy helped in producing 549.1 lakhs of Raily cocoons/ year and directly and indirectly helped the tribal community (Table-1) in creation of gainful employment and creation of sustainable income. This created awareness for forest conservation thereby maintaining environmental balance (Fig.1). Thus preservation of food plants of wild silkworms in their natural habitats along with preservation of the niche / habitat will promote interaction of the biotic elements and reduce environmental degradation.

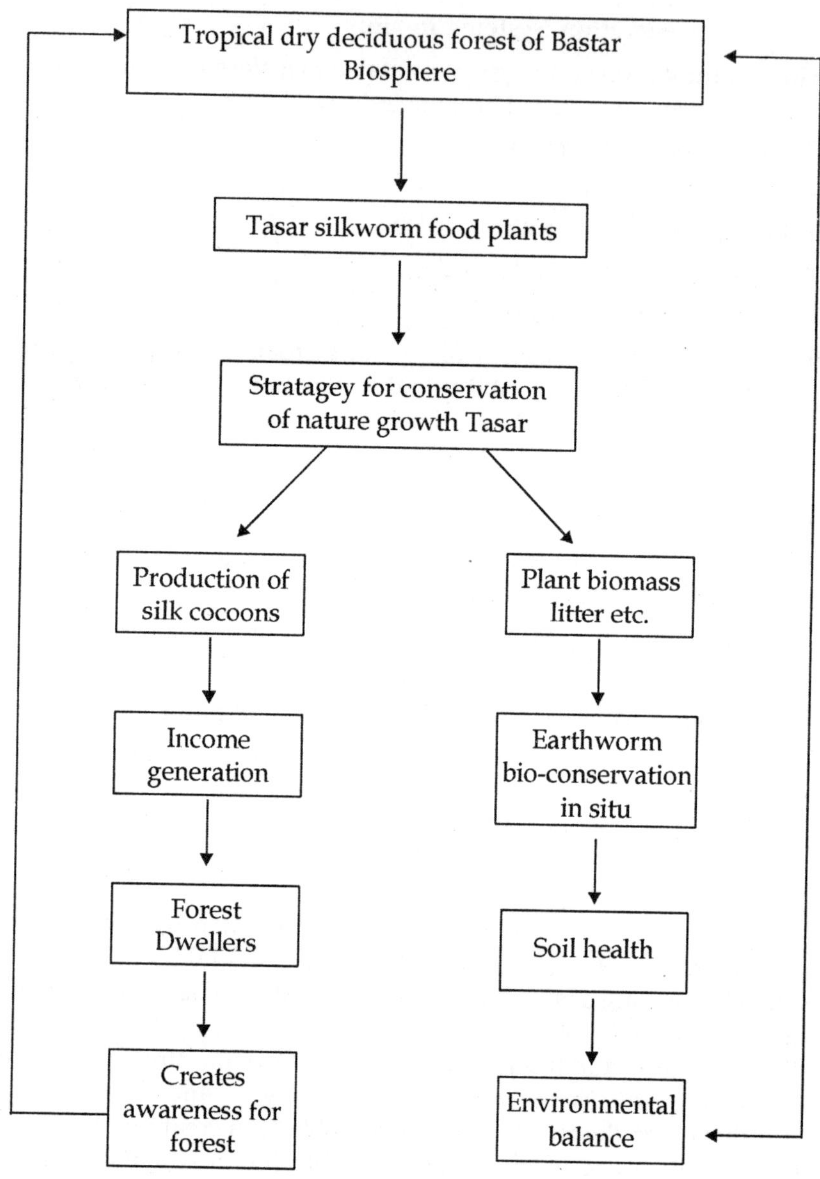

Fig. 1 : Inter dependence of Baster silkworm and earthworm in a forest ecosystem for balancing of the environment.

Table 1. Impact of RAILY development project on community developement.

Year	Production of			No. of man days created (in lakh)			Total man days during the year (in lakh)	Income generated (Rs. in crore)			Total income generated in a year
	Cocoons (in lakh)	Yarn (MT)	Farbric (lakh mtr.	At cocoon level	At yarn level	At fabric level		At cocoon level	At yarn level	At fabric level	
1997-98	576.30	9.00	18	11.88	9.00	9.00	29.88	9.22	10.80	36	56.02
1998-99	458.00	7.15	14	9.45	7.15	7.00	23.6	3.66	8.58	28	40.24
1999-00	513.00	8.01	16	10.53	8.01	8.00	26.54	4.10	9.61	32	45.71
2000-01	496.00	7.75	15	10.26	7.75	7.50	25.51	3.97	9.30	31	44.27
2001-02	570.42	8.90	17	11.88	8.90	8.50	29.28	5.70	10.68	35	51.38
2002-03	420.00	6.56	13	8.64	6.56	6.50	21.7	5.25	7.87	26	39.12
2003-04	810.0	12.65	25	16.78	12.56	12.50	41.84	10.12	15.18	50	75.30
Total average	3843.72 549.10	60.02 8.57	118 16.86	79.42 11.35	59.93 8.56	59.00 8.43	198.35 28.34	42.02 6.00	72.02 10.29	238.00 34.00	352.04 50.29

Interdependency of tasar silkworm and earthworm

Raily tasar silkworm propagation in the forest ecosystem excretes litter in the form of pellets. These excreta along with plant biomass form the feeding material of the earthworms and converts into in situ vermicompost. This organic material provides nutrition to plants and enhances soil health. This bio-geo-chemical cycle is very significant in a forest ecosystem (Fig.1). Protection of biodiversity is a useful capital, which not only saves the species but also the presence of genes. Ecological dependence of biotic elements is the main stay in any sustainable environment. The co-existence of mankind with nature in Bastar Biosphere Reserve will be an asset for economic and environmental protection.

PART - IV

EARTHWORMS IN WASTE MANAGEMENT

Chapter **10**

Vermi-composting of Municipal (Organic) Solid Waste and its Implications

M. Vikram Reddy and Swati Pattnaik
Department of Ecology and Environmental Sciences, Pondicherry Central University, Puducherry -605 014 (India).

The rapid population growth and economic development in the country and continuous human migration from country-side to urban municipalities and corporations has lead urban population reaching about 28% of the India's population being distributed across different towns, cities and metropolis. The waste generated in these urban conglomerations, which depends on the changing life style and living status including food habits, and is of great concern as its daily per capita production is of about 0.49 kg, i.e., half a kg, is showing a gradual temporal and spatial increase (0.456 kg in 1995 increased to 0.468 kg in 1997). The used and left-over materials comprising of household garbage and rubbish, street sweeping, construction and demolition debris, sanitation residues, trade and industrial refuse and bio-medical solid waste from various sources such as domestic, market, commercial and industrial sources, are collectively known as Municipal Solid Waste (MSW) and pose serious management problems. The amount of MSW generated in India is about 0.1 million t/day, which is approximately 36.5 m t/year. There has been a significant increase in MSW generation in India. Of the total MSW, the average collection ranges from 50% to 90% most of which is dumped out on land unscientifically and about 5% is channelled for composting (CPCB, 2000). The expected generation of MSW by 2025 will therefore be around 700 g per capita per day which is estimated to reach >260 million tons annually by 2050, more than five times the present level and need an additional 1400 km^2 area for its disposal, most of it in urban areas (TERI, New Delhi).

Some other estimates revealed that urban India produces about 42.0 million tons of MSW annually i.e. 1.15 lakh metric tons per day (tpd), out of which 83,378 tpd is generated in Class-I cities. The MSW generated in these cities works out to 72.5% of the total waste generated each day and this needs to be tackled on priority basis. It is 21,100 tpd (18.35%) in six mega cities; 19,643 tpd (17.08%) in metro cities having population 10 lakhs plus; 42,635.28 tpd (37.07%) in other Class-I towns with 1.0 lakh plus population. The details of MSW generated in various urban systems starting from some of the metropolis to towns are as follows:- New Delhi, the nation's capital producing 7,700 tpd; Chennai - 3,500 tpd; Bangalore - 2,000 to 2,300 tpd; Hyderabad- 2,200 tpd; Visakhapatnam- 1,000 tpd; Vijayawada - 450 tpd; Pune - 900 tpd; Madurai - 400 to 450 tpd; Pondicherry - 160 tpd; and Vellore - 60 to 80 tpd. Considering that the urban population of India is expected to grow up to 45% of total (World Bank) from the prevailing 28%, the magnitude of problem is likely to grow to even larger proportions unless immediate steps are taken to control waste generation, ensuring better collection and disposal leading to its sustainable management.

The "National Solid Waste Association of India (NSWAI)" has been formed a decade back, on 25th January 1996 in order to augment, operate and maintain MSW systems in a sustainable manner by urban local bodies, which should be cost effective and environmentally sound to prevent urban waste from causing environmental pollution and health hazards. The association is also a member of the International Solid Waste Association (ISWA), and provides forum for exchange of information and expertise in the field of MSW at the international level. The management of MSW has some basic components such as segregation, collection, transportation and disposal. The first most important step in solid waste management is segregation. The MSW need to be separated from recyclable waste like plastic, paper, glass leather and inert debris, and should be put in separate containers. This would help in proper disposal of the solid waste. The collection methods currently adopted by urban local bodies (ULBs) in India are primitive and lack specific standards or guidelines. Although the Manual on Municipal Solid Waste Management (2000) has taken a positive step in this direction, much still remains to be done. The Manual also identifies lack of technical, managerial, administrative and financial resources, and weak institutional structure of ULBs as the prime reason for poor MSW management in urban India. The problem of MSW management is notable not only because of its enormous quantity and volume, but also its spatial spread across 5161 cities and towns, and a variety of the problems involved in setting up managing systems for collection, segregation, transportation and disposal of the waste. The MSW broadly comprises organic (biodegradable) and non-organic (non-biodegradable) wastes including recyclable materials, which could be sorted out and reused as raw materials that generate income for the rag pickers. The organic fraction or the biodegradable matter of MSW ranges from 30 to 55% (some other reports state it as 60 to 70 %) and causes wide varieties of environmental health problems. It can be profitably converted into useful end products like compost and vermicompost (organic manure), methane gas (bio-methanation) (for cooking, heating, lighting, production of energy) etc. through composting including vermi-composting and production of energy including Refuse Derived Fuel (RDF) / pelletization, incineration and pyrolysis / plasma gasification.

The quantity and quality of MSW vary according to the socio-economic status and cultural habits, prevailing climate, location, urban structure, density of population and extent of non-residential activities (CPCB, 2000). Nevertheless, various technology options are available for processing the MSW. The objective of MSW management is to reduce the quantity of solid waste dumped on land by recovery of materials and energy from the waste in a cost effective and environment friendly manner. However, the introduction of new materials, especially packaging materials, plastics and the like pose a different set of disposal problems due to their inherent non bio-degradability.

Adverse effects of incorrect disposal of MSW

According to a CPCB study, about 94 % of the cities resort to indiscriminate and unscientific dumping of MSW. The un-engineered dump sites permit fine organic matter mixed with percolating rainwater to form leachate. The potential for this leachate to pollute adjoining water and soil is high, resulting in contamination of ground as well as surface water causing **water** pollution: i) it causes not only leaching of various nutrients - N, P and K but also a variety of heavy metals like - Cd, Pb, Cr, etc. to the deeper layer of soil, causing **groundwater** pollution of nitrates and total dissolved salts (TDS) and heavy metals; ii) ammonia toxicity; and iii) nitrate and phosphate enrichment of surface water causing 'eutrophication'.

Burning of MSW in dumping yards leads to **air** pollution contaminating the air with dioxins and furans; that's why the burning of MSW is banned in many cities like New Delhi. i) It causes thick clouds of smoke accompanied by pungent odor/stench and dropping visibility to a minimum, causing eye burning and irritation; ii) release of ash particles and toxic gases to the air causing lung and skin related illness also; and iii) the emit ion of dioxin and furan (polychlorinated compounds) cause cancer, breathing-related diseases, and may retard tissue growth in the brain. Dumping of MSW also causes soil pollution: i) elevated nutrient inputs of N, P, K, Mg and Ca to the surface soil; ii) ammonia toxicity to soil and its beneficial soil biota; iii) increase in heavy metals in the upper soil layers; iv) after long-term use increases soil pH and v) anaerobic and anoxic decomposition emitting malodor vi) spreading bacterial, viral and other parasitic diseases.

The sanitary landfill based disposal system does not seem to be sustainable largely as land is becoming scarce in urban areas besides the implications of this method of disposal for pollution of ground water and air pollution due to release of Greenhouse Gases (GHGs). The degradable organic matter from these wastes when dumped in open undergoes either aerobic or anaerobic degradation or forms the breeding grounds for **pathogens** and vectors such as insects and other animal vectors like rodents; swarms of mosquitoes, house-flies and rats that caused worst epidemics like Plague in Surat.

The cities are therefore looking, for a system of disposing the solid waste that could be free from hazards as well as cost effective. Concerns for environmental risks have driven some of the cities to adopt newer ways of waste disposal that are neither hazardous nor unaffordable. The two leading innovative mechanisms of waste disposal being adopted in Indian urban systems like Pune include composting (aerobic composting and vermi-composting, etc.) and waste-to-energy (bio-methanation, and pelletisation, pyrolysis/ gasification) in cities like Hyderabad. India where a lot of solid organic waste is available in different sectors with no dearth of manpower, the environmentally acceptable vermi composting technology using earthworms can very well be adopted for converting organic waste into wealth. The viability of using earthworms as a treatment or management technique for numerous organic waste streams has been investigated and is followed. The present paper aims at providing an overview of various aspects of vermicompost of MSW and its potential implications.

Vermi-composting of MSW

It is an environmental-friendly solution for garbage disposal in urban areas. In this process, earthworms are used in bioconversion of organic portion of MSW into an amorphous dark colloidal humus substance known as 'Vermi compost', under condition of optimum temperature, moisture and aeration. According to Environmental Protection Agency (USA), 'Vermi composting is the bioconversion of organic waste materials through earthworm consumption' (EPA, 1980). The renowned naturalist Charles Darwin was first to carry out observations scientifically on the role of earthworms in vegetable mould formation a century back. Since then, researchers all over the world conducted investigations on vermi composting. Darwin's scientific studies on earthworms spread over four decades were published (Darwin, 1881), which was celebrated as 'Centenary celebration' in 1981 at Grange-over-Sands in Cumbria (United Kingdom) by Professor. J. E. Satchell organizing the First International Symposium on Earthworm Ecology (ISEE-I). It was attended by a few Indian earthworm-researchers (including myself - MVR), who were informed and educated on the concepts of 'Vermi-composting' of MSW by the participants from other countries viz., Ms. David Livingston from California, USA and Professor M. Bouche from France; thus, the technology has been brought to India and is about 25 years old in this country.

Vermi composting, a vermi-technology or earthworm technology involves large scale application of earthworms belonging to twenty different species found in India but only a few, five of them are used widely in vermi composting of solid organic wastes (ZSI, 1993; Reddy, 2002). They are *viz., Perionyx excavatus* (Perrier, 1872), *Lampito mauritii* (Kinberg) and *Polypheretima elongata* (Erseus) belonging Megascolecidae; *Eudrilus eugeniae* (Kinberg, 1867) belonging to Eudrilidae; *Eisenia fetida* (Savigny, 1826) belonging to Lumbricidae, the former three species are local and the later two are exotic. They process the organic waste to produce value added vermi-compost and vermi-protein keeping the environment clean and green reducing environmental pollution. Vermi composting technology is safer in terms of risks to human and environmental health, compared to direct application of municipal solid organic waste (MSOW) in agricultural fields, as practiced in many places. Composting using earthworms is about 25% faster in decomposition of organic waste, as these worms along with meso-fauna particularly different species of Acarina and

Collembola and microbes take part in devouring and digesting the waste producing a better end product i.e., vermicastings, than the mere compost that is solely of microbial origin (Department of Natural Resources (DNR) 2001; Panikkar and Riley, 2003).

Conversion of urban organic solid wastes into vermicompost can be done at two stages; one at the domestic level by each household and the other at larger vermicompost units to biodegrade the garbage of a town or city or locality. At the domestic level, earthworms can be cultured even directly adding them to the containers used for depositing organic wastes from kitchens and households, where the earthworms can convert the waste into vermicompost. Organic solid wastes from different localities of a city can also be converted into vermicompost in larger units. Some organizations in Tamil Nadu, Delhi and Himachal Pradesh are engaged in degrading the city garbage into useful organic manure. In countries like Cuba, the vermi composting programme started at national level during 1986, and by 1992 there were 172 vermicomposting centers producing 93,000 tons of vermicompost that boosted the agriculture productivity of the country.

Action of verms (earthworms)

The earthworm action is both physical and biochemical during the vermi-composting process. The physical process includes substrate aeration, mixing as well as grinding in the gizzard. The biochemical process is influenced by microbial decomposition of the substrate including the digestion in the intestine of earthworms, and during the process while the devoured organic matter moves in the gut it is added with microbial metabolites that are the phyto-hormonal like substances (Hand *et al.*, 1988), with the help of enzymes like proteases, lipases, amylases, cellulases and chitinases which bring about rapid biochemical changes of the cellulosic and the proteinaceous materials of the substrate.

Earthworms have been known to secrete certain fluids of anti-bacterial properties (Satchell, 1983). They release coelomic fluids on the decaying waste biomass, which have an anti-bacterial property and kill the pathogens and thus, reduce not only the pollution but also minimize the harmful effects of wastes. Most human pathogens, including *Escherichia coli* and *Salmonella* are killed by earthworm's activity. More studies are required on this aspect to shed light on the processes that attain pathogen reduction and removal. Competition between pathogenic organisms and indigenous micro-flora in the earthworm castings or vermicomposting for nutrients has been shown to cause pathogen suppression and reduction (Eastman et al., 2001; Ellery, 2000; Morgan and Morgan, 1999). The microbes present in earthworm casts/compost have deodoring effect (Lee, 1985). Vermicomposting converts organic wastes into stable environment-friendly rich manure with potentially high economic value. Vermicompost is a pathogen free product and rich in microbial diversity, population, and activity (Subler *et al.*, 1998). Earthworms along with microbes have the ability of decomposing lignin and accelerating humification process and manage the wastes in such a way that the leaching of harmful pollutants into the deeper layer of ground is prevented and hence protect groundwater from pollution.

Other advantages of vermi-composting

i). Earthworms play a pivotal role in serving as versatile natural bioreactor, converting organic wastes into valuable organic manure. Vermi-composting process helps in the management of MS (O) W, reducing and stabilizing its bulk density (up to 70 per cent).

ii). During the process they can accumulate certain persistent heavy metals, and pesticides residues. It tends to prevent the amounts of bio-available heavy metals present in MS (O) W.

iii). They also break down complex bio-molecules present in the garbage into simple compounds that are utilized by the micro-organisms releasing useful metabolic compounds like plant growth hormones, antibiotics, vitamins, etc. into casts/excreta.

iv). They not only lower pathogenic microbes but miraculously eliminate the foul-smelling malodor, abating organic pollution, and enhance the useful microbes like *Azotobacter* sp. and actinomycetes.

v). Vermicompost is a stable non-toxic material that has economic and ecological value as soil-conditioner for plant growth.

vi). It acts as complex fertilizer granules synchronizing and supplying a suitable mineral nutrient balance.

vii). While it acts as a biocide against diseases, suppresses plant pathogens; and in turn help in reducing the pesticide pollution in the environment.

viii). It acts as porous and aerated material with excellent moisture holding capacity.

ix). Eco-farming in China with earthworms known as "Green profits-earthworm excrement" i.e., vermi-compost when applied as manure to paddy fields in some areas of Zhejiang province in China, increased the yield of rice reaching over 15,000 kg per hectare, without use of any other fertilizer.

Soil amendment with vermi compost

Vermi compost also improves structural stability of soil stimulating the formation of soil aggregates through bacterial polysaccharides by enhancing the rate of organic matter degradation. Researchers analyzing the chemical composition of vermi compost have reported reductions in pH, EC, C/N and C/P ratio (Elvira *et al.*, 1998) and increase in total N, P, and K (Suthar, 2007) and available N, P, K, Ca, Mg, Mn, Na, Fe, Cu and Zn (Edwards, 1998; Bansal and Kapoor, 2000). MSW compost and vermicompost also include that above nutrients, and improves the soil physical properties, nutrient retention capacity and stimulate the microbial activity; thus, may improve plant growth and decrease the leaching of the pollutants into surface- and ground- water.

Plant nutrients of Vermicompost :

During the process of vermicomposting, the nutrients present in the wastes in complex form are converted into the simpler forms that are more readily absorbed by the plants. The surface area of vermi compost particles provides many micro-sites for microbial activity and for the strong retention of nutrients. The nutrient status of vermicompost produced with different organic waste is: organic carbon 9.15 to 17.98 %, total nitrogen 0.5 to 1.5 %, available phosphorus 0.1 to 0.3 %, available potassium 0.15, calcium and magnesium 22.70 to 70 mg/100g, copper 2 to 9.3 ppm, Zinc 5.7 to 11.5 ppm and available sulphur, 128 to 548 ppm (Kale, 1995). A standard vermi compost prepared from organic solid waste of market waste, but not of the MSW, is characterized with pH, EC and organic carbon are 7.72; 0.38 and 18.81, respectively and the total nitrogen, total phosphorus and total potassium of 1.34, 1.62 and 1.15 % respectively (Ansari and Ismail, 2001). Our findings showed variations in concentrations of N, P, K, Ca and Mg in vermi compost of different species of earthworms prepared from MS (O) W, and the nutrients being higher in the vermi compost compared that of sole compost and the substrates, MSW (Swati and Reddy, 2007).

Vermi compost as organic fertilizer

Vermicompost contains even micronutrients and some additional substances that are often lacking in commercial fertilizers, so compost particularly the vermicompost is an essential dietary supplement for any soil. It is slightly alkaline with more soluble salts, exchangeable sodium, potassium and magnesium (Singh and Sharma, 2002). Due to the variable content and slow release of major and minor nutrients, vermicompost is often considered a supplement to fertilizers and is a readily available nutrient source. When applied continuously over a period of time, vermicompost may provide significant amount of the nutrients and minimize the application of additional fertilizers.

The plant growth regulators and other plant growth influencing substances acting like auxins, cytokinins, gibberellins, and that humic nature, produced through microbial metabolism in the earthworm gut have been reported to be present in the vermicompost (Atiyeh *et al.*, 2002). The humic materials extracted from vermicompost have been reported to produce auxin- like effect on cell growth and nitrate metabolism in carrot (*Daucus carota*) (Muscolo *et al.*, 1999). However humic substances can also occur naturally in animal manure, sewage sludge or paper - mill sludge but their amount and rate of production are increased dramatically by vermicomposting. The vermicompost has been reported to contain a higher base exchange capacity and rich in total organic matter, phosphorus, potassium and calcium that increase the oxidation potential and reduce the water-soluble chemicals that constitute possible environmental contaminants. It contains enzymes such as protease, amylase, lipase, cellulase and chitinase that continue to disintegrate organic matter even after they have been ejected. It is considered as an excellent product since it is homogenous and has reduced level of contaminants tending to hold more nutrients over a longer period without causing any impact on the environment (Ndegwa and Thompson, 2001).

Potential effects of vermi compost on plant growth

The vermicompost has been found to be ideal organic manure for boosting the plant growth and enhancing biomass production of a wide variety of crops (Edwards, 1998). Galli et al. (1992) reported an increase of 30% in protein synthesis in seedling of *Lactuca sativa* following the application of vermicompost. The compost produced by earthworms from municipal wastes (CPEMW) caused a reduction in the soil pH and increased the dry matter production of maize

(Ferreira *et al.*, 1992). This increase was significant when CPEMW was used in combination with lime or mineral fertilizer or both. Earthworm castings at the rate of 4 t/ha replace 45 t/ha of cow manure formerly used and improved production by 31 % and enhanced the quality by reducing leaf chlorine content from 1% to 0.4 % (Werner, 1997). Vermicompost of grape processing waste when applied to vines under shadow mulches, increased grape yield by 20-50% at the first harvest in Australia (Edwards and Steele, 1997). The growth of plants with vermicompost application is also influenced by reduction of bio-available heavy metals and elimination of pathogens (Dominguez *et al.*, 1997). Vermicompost was used as bio-pesticides in tomatoes and cabbages in Poland, which protected the vegetable plants from a number of microbial diseases (Szczech and Smolinska, 2001).

Vermi composting of MSW and global warming

Organic waste on dumping sites release the GHGs - Carbon dioxide (CO_2), Methane (CH_4) and Nitrous oxide (N_2O) that trap heat in the atmosphere adding to global warming; CH_4 absorbs 20 to 30 times as much infrared radiations and are accumulating in the atmosphere about twice as fast as CO_2. Proportions of global warming due to these gases are i). CO_2 - 64 %, ii). CH_4 - 19 % and iii). N_2O - 6 %. The earthworms prevent release of these harmful GHGs by devouring the organic waste converting it into useful vermicompost.

Heavy metal contamination of MSW and vermi compost

Of late, farmers are subsidized by the various Governments for producing and using vermi-compost/ compost prepared from organic wastes including the agriculture waste and MSW in various southern states such as Tamil Nadu, Pondicherry, Karnataka and Andhra Pradesh. However, adding the MSW vermi compost/compost to agriculture soils may act as a significant source of persistent heavy metals (Cd, Hg, Pb, As, Cr, Cu, Ni, Zn, etc.), as MSW contains these pollutants; the sources being batteries, consumer electronics, ceramics, light bulbs, paint chips, lead foils, wine bottle closures, some inks, plastics, etc. Some of these elements such as boron, zinc, copper and nickel are essential in small amounts for plant growth; but they decrease plant growth when present in higher amounts. The impact of metals on crop-plants grown on compost amended soils depends on their concentrations as well as soil conditions such as pH, Organic Carbon, etc. Soils cropped for many years may fall deficient of micronutrients such as Boron, Zinc and Copper, and soil amendment with the MSW compost could mitigate such deficiencies. Heavy metals such as Cd, Pb, Hg and As are of concern because of their potential harm not only to soil and soil organisms but also to animals and human beings.

Animal and human health

Heavy metals like Cd (0.01-ppm max. desirable limit)), Pb (0.01 ppm) and Hg (0.001 ppm) are very harmful to animals and human beings even at relatively low concentrations. Cadnium content of crops particularly in Tobacco is of concern because when tobacco is burnt and inhaled, much of the Cd is taken up by the human body. Leafy crops like spinach are known to accumulate Cd (MPL – 0.5 mg/kg or less). The risk of heavy metal movement from the substrate - MSW to the crop-plants is higher. These metal pollutants though present in negligible concentrations may pass from the MSW to the compost/ vermi-compost and when the latter is used in crop fields these pollutants move to the edible plant-parts, and enter the food-chain affecting environmental quality and finally human health causing various afflictions. This has raised concerns for protection of environment and human health that are important for sustainable development.

Heavy metals usually present at levels as low as ppb in MSW dumping sites, accumulate in biological system in many folds. Earthworms though play a significant role in detoxifying the MSW and the contaminated sites; the low-cost vermicomposting technique can be used in the reduction and removal of toxic metals from MSW (Jain *et al.*, 2004). Earthworms can take up and accumulate heavy metals such as Cd, Hg and Au etc. in their tissues to a certain level, when living in contaminated environments (Hughes *et al.*, 1980; Beyer, 1981). The concentration of the heavy metals in earthworm tissue depends on the source of metals and species of earthworms. They can survive by metal detoxification involving binding and storage of the metals in metallothionein and metal-binding proteins (Dallinger, 2000).

The heavy metals present in the environment form complex aggregates with the humic substances and the polymerized organic fractions of solid waste. These toxic complex aggregates while passing through earthworm gut are split into simple molecules of less toxic nature due to action of digestive enzymes. The total amount of heavy metals are increased by 25-30 % as a consequence of the carbon loss by mineralization while passing through the gut; however, the amount of bioavailability of the metals tends to decrease by 35-55 % because of the chemical binding of the metals with metal binding proteins present in worm's gut (Dominguez, 2004). The earthworm chloragosomes aid in detoxifying the pollutants, which consist of modified epithelial cells around the gut containing constituents of ion exchange compounds - phosphoric acid, carboxyl, phenolic hydroxyl and sulphonic acid groups. The chloragosomes act as a cation exchange system capable of taking up and retaining heavy metals, thus reducing the toxic effects (Ireland, 1978).

Non-enzymatic metalloproteins or metallothioneins present in the earthworm's gut also play a significant role in detoxification of heavy metal pollutants. These are associated with the cytoplasmic fraction of earthworm tissue exposed to high levels of heavy metals like cadmium, copper, gold, mercury, silver, zinc, etc. These protein granules are characterized by the low molecular weight and peak absorption of 250 nm, and are capable of binding with the toxic heavy metals converting them to less toxic compounds. The metal binding proteins including the cadmium binding protein have been identified in various earthworm species *viz., Eisenia fetida, Dendrodrilus rubidus, Aporrectodea caliginosa* and various *Lumbricus* species. Many zinc and calcium metallo-enzymes have been identified from earthworm tissue.

Conclusion

Vermicomposting technology involves harnessing earthworms as versatile natural bioreactors, playing a vital role in the decomposition of a wide variety of organic waste including MS (O) W, which when applied to soil maintained its fertility and boosted plant growth. It reduced pollution burden of the environment keeping it clean as well as green. Hence, using earthworms in producing vermicompost of organic wastes has a good scope for developing it as a municipal solid waste management technology in developing countries like India, which reduce environmental pollution and enhance the crop productivity.

Acknowledgements

Swati Pattnaik is thankful to Pondicherry University (Puducherry) and UGC (New Delhi) for the award of a research fellowship.

References

Atiyeh, R.M., Lee, S., Edwards, C.A., Arancon, N.Q. and Metzger, J.D. 2002: The influence of humic acids derived from earthworm-processed organic wastes on plant growth. *Bioresource Technology*, 84, 7-14.

Ansari, A.A. and Ismail, S.A. 2001: Vermitechnology in Organic Solid Waste Management, J. *Soil Biol. Ecol.*, 21, 1 & 2, 21-24.

Beyer, W.H. 1981: Metals and terrestrial earthworms (Annelids: Oligochaeta), In: *Workshop on the Role of Earthworms in the Stabilization of Organic Residues* (M. Appelhof, Compiler), Proceedings, Beech Leaf Press, Kalamazoo, MI, p.137-150.

Bansal, S. and Kapoor, K.K. 2000: Vermicomposting of crop residue and cattle dung with *Eisenia fetida*, Bioresource *Technology*, 73, 95-98.

CPCB (Central Pollution Control Board), 2000: Management of Municipal Solid Waste, Parivesh Newsletter of Central Pollution Control Board, Delhi.

Dallinger, R. 2000: Metal tolerance and metabolism in terrestrial invertebrates: molecular and biochemical aspects, *Comparative Biochemistry and Physiology - Part A: Molecular & Integrative Physiology*, 126, 1, 36.

Darwin, C.1881: *The formation of vegetable mould through the action of worms, with observations on their habitats.* Murray, London, p.326.

Department of Natural Resources (DNR), 2001: 'Worm composting system', Department of Natural Resources, Missouri, USA, <www.dnr.state.mo.us> Accessed 24 July 2001.

Dominguez J., Edwards, C.A. and Subler, S. 1997:Comparison of vermicomposting and composting. *Biocycle*. 38, (4): 57-59.

Domínguez, J. 2004: State of the art and new perspectives in vermicomposting research. In: C.A. Edwards (Ed.), *Earthworm Ecology*, 2nd Ed, CRC Press, Boca Raton, p. 401–425.

Eastman, B.R., Kane, P.N., Edwards, C.A., Trytek, L., Gunadi, B., Sterman, A.L. and Mobley, J.R. 2001:'The effectiveness of Vermiculture in human pathogen reduction for USEPA Biosolids stabilization, *Compost Science & Utilization*, 9, 1, 38-49.

Edwards, C. A. and Steele, J. 1997: Using Earthworm systems, *Biocycle*, 38 (7), 63-64.

Edwards CA. 1998: The use of earthworms in the break down and management of organic wastes. In: *Earthworm Ecology*. CRC Press LLC, Boca Raton FL, p. 327-354.

Ellery, D. 2000: 'Earthworms make great waste managers', Solid waste online: 32-8-00,www.solidwaste.com/content/news/> Accessed 30 July 2001.

Elvira, C., Sampedro, L., Benítez, E. and Nogales, R. 1998: Vermicomposting of sludges from paper mill and dairy industries with *Eisenia andrei*: a pilot-scale study, *Bioresource Technology*, 63, 205-211.

E. P. A., 1980: Compendium of Solid Waste Management by Vermicomposting (US Environmental Protection Agency, Ohio), p.45.

Ferreira, M.E, Cruz, MCP, and da Da Cruz, MCP. 1992: Effects of compost from municipal wastes digested by earthworms on the dry matter production of maize and on soil properties, *Cientifica Jaboticabal*, 20, 1, 217-226.

Galli, E., Rosique, J.C., Tomati, D., Roig, A., Senesi, N., and Miano, T.M. 1992: Effect of humified materials on plant metabolism. In: *Humic Substances in the Global Environment and Implications on Human Health.* Proceedings of the 6th International Meeting of the International Humic Substances Society, Monopoli (Bari), Italy, p.595-600.

Hand, P., Hayes, W.A., Satchell, J.E., Frankland, J.C., Edwards, C.A. and Neuhauser, E.F. 1988: The vermicomposting of cow slurry, *Earthworms in Waste and Environmental Management*, p. 49-63.

Hughes, M. K., Lepp, N.W. and Phipps, D.A. 1980: Aerial heavy metal pollution and terrestrial ecosystems, In: *Advances in Ecological Research* (A. Macfadyen, Ed.), Academic Press, London, p.218-327.

Ireland, M.P. 1978: Heavy metal binding properties of earthworm chloragosomes, *Acta Biol. Acad. Sci. Hung.*, 29, 385-394.

Jain, K., Singh, J., Chauhan, L. K. S, Murthy, R. C. and Gupta, S .K. 2004: Modulation of flyash-induced genotoxicity in *vicia faba* by vermicomposting, *Ecotoxicology and Environmental Safety*, 59, 1, 89-94.

Kale, R.D. 1995:Vermicomposting has a bright scope, Indian Silk, 34,6-9.

Lee, K.E. 1985: *Earthworms: Their ecology and relationships with soils and land use.* Academic Press, Sydney, Australia, p.411.

Morgan, J.E. and Morgan, A.J. 1999: The accumulation of metals (Cd, Cu, Pb, Zn and Ca) by two ecologically contrasting earthworm species (*Lumbricus rubellus* and *Aporrectodea caliginosa*): implications for ecotoxicological testing. *Applied Soil Ecology*, 13,9-20.

Muscolo, A., Bovolo, F., Gionfriddo, F. and Nardi, S. 1999: Earthworm humic matter produces auxins like effect on *Daucus carota* cell growth and nitrate metabolism. *Soil Biol. and Biochem.* 31,1303-1311.

Ndegwa, P.M. and Thompson, S.A. 2001: Integrating composting and vermicomposting in the treatment of bioconversion of biosolids. *Biores. Technol.*, 76, 107-112.

Panikkar, A. and Riley, S. 2003: Biological treatment of black water and mixed solid waste: An example of a treatment system, In: *Proceedings of Orbit 2003*, Vol.2, Murdoch University, Perth, Australia.

Reddy, M.V. 2002: Vermicomposting - A latent technology to improve the productivity of small farmers, In: *Rural technology for Poverty Alleviation*, (Ed. P. Purushotham), National Institute of Rural Development, Hyderabad, p. 46-55.

Satchell, J.E. 1983: Earthworm microbiology, In: *Earthworm Ecology: From Darwin to Vermiculture*, J.E. Satchell (Ed.), Chapman and Hall, Cambridge, UK.

Singh, A. and Sharma, S. 2002: Composting of a crop residue through treatment with microorganisms and subsequent vermicomposting, *Bioresource Technology*, 85, 107-111.

Subler, S., Edwards, C.A. and Metzger, J.D. 1998: Comparing vermicomposts and composts, *Biocycle*, 39, 7, 63-66.

Suthar Surendra, 2007:Nutrient changes and biodynamics of epigeic earthworm *Perionyx excavatus* (Perrier) during recycling of some agriculture wastes, *Bioresource Technology*, 98, 8, 1608-1614.

Swati, P. and Reddy, M.V. 2007:Major nutrients of compost and vermicompost proceed by *Eudrilus eugeniae* (Kinberg), *Eisenia fetida* (Savigny) and *Perionyx excavatus* (Perrier) - I. Municipal (organic) Solid Wastes, *Biol. Fertil. Soils* (communicated).

Szczech, M. and Smolinska, U. 2001: Comparison of suppressiveness of vermicomposts produced from animal manures and sewage sludge against *Phytophthora nicotianae* Breda de Haan var. nicotiannae. *J. Phytopathology*, 149, 77-82.

Werner, M.1997: Earthworm team up with yard trimmings in orchards, *Biocycle* 38, 64-65.

Zoological Survey of India, 1993: Earthworm Resources and Vermiculture, Z. S. I., Calcutta, p.128.

Chapter 11

Composting of Paper Mill Sludge using Native Earthworms and Impact on the Growth and Yield Parameters of Black gram, *Vigna mungo*

S.Umamaheswari, V. Balamurugan and G.S.Vijayalakshmi
Sri Paramakalyani Centre for Environmental Sciences, Manonmaniam Sundaranar University, Alwarkurichi - 627 412

Waste is the resource, which is left unused, and it gets accumulated due to the factors viz., urbanization, increase in human population and industrialization. The waste management licensing regulations (HM Government, 1994) allow certain non-agricultural wastes to be applied to agricultural land without the need for a disposal license where this activity results in a "benefit to agriculture or ecological improvement". The wastes, to which this exemption from licensing applies, include paper mill sludge, waste paper and de-inked, paper mill sludge. Pulp and paper industry are one of the largest consumers of water, consuming about 250-350 m^3 of water/ ton of paper produced and generates about 225-275 m^3 of waste water/ ton of paper and produce large volumes of sludge (Raghuveer and Sastry, 1995). The paper mill sludge, the cellulosic waste introduced into the water body further degrades more slowly thus increasing the BOD level of the water body.

Physical, chemical and biological processes bring about treatment of wastewater. The physical and chemical process although can be designed for any degree of purification are highly expensive (Gohil, 1995). Dehydration of the sludge to cake and discarding off in landfill sites or incineration would lead to the loss of a profitable cellulose based resource for which biological treatment methods have received much attention. Vermiculture biotechnology is a dynamic process brought about by the earthworms with the aid of mixed microbial populations. Worms ingest the solids and convert a portion of the organics into worm biomass and to respiration products and expel the remaining partially stabilized matter, which is referred to as "worm casts" rich in essential micro and macronutrients required for the plant growth. Worm castings is a high quality humic product, which can be used as soil amendment (Ceccanti and Masciandaro, 1999). Microorganisms play an important role in decomposition of soil organic matter, which mainly comprises of lignocellulose. Fungi are known to be active degraders of organic matter in soils (Alexander, 1978).

In the present study, the paper mill sludge, raw material as solid biodegradable effluent was utilized to convert into a utilizable biofertilizer through the application of earthworms, *Lampito mauritii*.

Sludge collection and vermicomposting

Paper mill sludge was procured from the Sun Paper Mill (P) Ltd., Cheranmahadevi, Tamil Nadu and the cow dung standard bedding material, was collected from the nearby locality. The paper mill sludge was shred into small pieces and exposed to the sun to remove foul smell, transferred to shade, dried and used as substrates for composting.

The experimental bed was prepared with cow dung and sludge in 1:1 ratio in rectangular plastic tubs (of 12"x17"x51" size) in triplicates. Pre-weighed breeders of the indigenous earthworm species, *Lampito mauritii* each numbering fifteen were manually introduced in the tub. Control was maintained without earthworms. The experimental tubs were placed indoor in the laboratory to avoid direct sunlight, rain and to protect the worms from predators. Regular sprinkling of water on alternate days maintained optimal moisture of the beds at 50-55%. Regular mixing of the vermibeds was carried out without damaging the earthworms for homogenous decomposition. The average wet biomass of the earthworms and the number of cocoons produced by the earthworms were counted at every fortnight interval and maintained separately in 1:1 concentration. Hatching percentage of the cocoons were determined by separate maintenance of ten numbers and 5 cocoons were maintained separately in Petri plates containing sludge and cow dung in 1:1 ratio to obtain the number of hatchlings per cocoon.

The bedding material was analyzed for the physical parameters like pH, electrical conductivity, porosity and water holding capacity and the chemical parameters viz., nitrogen, phosphorus, potassium, calcium, magnesium, iron, organic carbon, C/N and C/P. Enumeration of microbial population was carried out by adopting the methodology described by Allen (1953).

The impact of vermi compost on the morphological and yield characteristics of black gram (*Vigna mungo*) with different combinations viz., sand + soil (1:1), sand + soil + Farm Yard Manure (1:1.5:0.5), sand + soil + FYM + BF (1:1.5:0.5+BF), sand + soil + Vermicompost (1:1.5:0.5), sand + soil + VC + BF (1:1.5:0.5+BF), sand + soil + VC + FYM (1:1.5:0.25:0.25) and sand + soil + VC + FYM + BF (1:1.5:0.25:0.25+BF). The plants were grown for a period of 60 days and five plants were harvested gently from each treatment and observed for the morphological characteristics viz., root and shoot length, number of leaves, leaf area index, and number of nodules, flowers, pods and seeds.

Experimental findings

Table-1. Biomass of native earthworms and their cumulative cocoon production in paper mill sludge

Parameter/ days	0	15	30	45	60
Biomass	0.628±0.01	0.660 ±0.05	0.769±0.02	0.857±0.05	0.915±0.03
Cocoon Production		5.33±2.85	9.0±2.52	13.0±4.04	5.33±1.45

Table 1 indicates the biomass of *Lampito mauritii* in 1:1 paper mill sludge and cow dung. No mortality was observed during the experimental period, a noteworthy feature of the paper mill sludge. The worms gained weight throughout the experimental period and the maximum weight was recorded at the end of the experiment. The maximum weight gain in different organic wastes was reported by Haimi and Huhta (1986) inactivated sludge mixed with or embedded in sieved pine bark. Addition of organic matter has resulted in increased population density of worm, biomass, casts as observed by Tiwari (1993), which coincides with the present study. The number of cocoons laid increased up to 45th day and then decreased. Lee (1985) found that the cocoons may be produced at any time of the year and the cocoon production is influenced by a variety of characteristics of population, mainly population density, biomass and by external factors especially soil temperature, moisture and energy content of available food. The varied number of cocoons produced in the present study might be due to the palatability of the feed as supported by Guild (1948).

Table-2. Fecundity status of the native earthworm in paper mill sludge

Hatching %	80.33±2.67
No. of hatchlings/cocoon	1.33±0.33
Total no. of hatchlings	65.67±9.70

The hatching percentage, numbers of hatchlings per cocoon and the total number of hatchlings produced by the native earthworms in the paper mill sludge have been shown in Table-2. The hatching percentage was found to be 80 for which the favourable temperature, moisture and other environmental factors might have paved the way for the hatchment of cocoons. The number of hatchlings per cocoon was recorded as 1.33. Hatanaka *et al.* (1983) stated that the number of hatchlings per cocoon varied from one to seven with an average of 3.9 in dairy waste sludge cake. The hatchling arrival, in general, depends on the preferential utilization of food by earthworm and the nature of substrates, which generally supported the reproductive behaviour of the earthworms. The native earthworms in the paper mill sludge produced total of 65 hatchlings.

Table-3. Physico-chemical characteristics of paper mill sludge before and after composting

Parameters	Initial	Compost	Final
pH	9.37±0.15	7.83±0.23	8.43±0.19
EC	0.74±0.03	1.17±0.05	0.94±0.07
Pore space	62.33±0.59	72.03±1.27	65.23±0.16
Water Holding Capacity	395.53±9.89	735.23±21.37	412.9±2.91
N	0.190±0.01	0.281±0.08	0.243±0.004
P	0.011±0.0004	0.177±0.004	0.014±0.002
K	0.095±0.01	0.185±0.02	0.099±0.01
Ca	0.441±0.006	1.469±0.05	0.671±0.007
C/N	229.0±4.64	45.49±3.18	173.8±4.45
C/P	4140.4±304.4	158.19±8.89	3170.47±506.56

The physico-chemical composition of the paper mill sludge mixed with cow dung in 1:1 is reported in Table 3. The pH was uniformly brought to a neutral level after composted by the native earthworms. This might be due to the production of CO_2 and organic acids by microbial activity during the process of bioconversion of the substrates in the bed. A gradual increase of EC was observed in the treatments including control from the initial value. The increase might be due to the presence of exchangeable calcium, magnesium and potassium in the worm cast than the soil (Bhatnagar and Palta, 1996). Pore space was observed to increase in all the treatments gradually during the experiment. An increase of pore space from 52.31% in the initial litter waste to 150.71% was observed in the *Lampito* compost by Umamaheswari *et al.* (2003a), which is in agreement with the present study. The value of WHC in the final compost was found to be more than that in the control. These findings corroborate with the finding of Umamaheswari *et al.* (2003b) in the mango litter waste composted by *Perionyx excavatus.*

The nitrogen values increased significantly over control after composting. This increasing trend in nitrogen in the vermibed is also reported by Balamurugan (2002). The phosphorus level increased in the present study in the earthworm treated compost on comparison with the initial and control samples. Graff (1971) recorded increased phosphorus in the casts than the surrounding soil, which coincides with our findings. The potassium level was observed to increase progressively more than the initial in the control and compost. This finding is in conformity of Jambhekar (1992). The increase of calcium in the vermicompost might be due to the excretion of calcium from the calciferous glands. The value of calcium at the end of compost formation was more than the control. Comparison of the initial and final values of the C/N and C/P ratios in the experiment showed a decreasing trend in the worm worked compost. This may be due to the mineralization of plant nutrients that derived organic material during the passage through earthworm with consequent low C : N ratios.

Table-4. Microbial population in the initial, control and worm worked beds of paper mill sludge

Microbe	Initial	Control	Compost
Bacteria (g^{-1} dry wt x 10^5)	38.67±1.76	59.33±0.88	123.67±2.40
Fungi (g^{-1} dry wt x 10^3)	79.67±1.2	232.67±4.18	297.67±7.22
Actinomycetes (g^{-1} dry wt x 10^4)	9.67±1.20	19.67±0.88	59.67±2.91

Enumeration of microbial population in the initial, control and worm worked compost beds have been depicted in Table-4. The microbial population was observed to be increased progressively in the control and worm inoculated beds compared to the initial samples. It is evident that the organic wastes used for the experimental study stimulated the microbial population during the passage through the gut of the earthworm. Preference of fungally contaminated cellulosic feed was demonstrated by Cooke (1983).

Table-5. Effect of vermi compost and soil mixture on the growth of black gram (in cm.)

Treatment	Root length	Shoot length	No. of leaves	Leaf Area Index
C	14.04±0.85	28.32±2.69	9.8±0.07	26.89±3.05
T1	15.24±1.09	31.96±3.57	10.4±1.8	35.34±3.46
T2	16.16±0.69	33.3±2.87	10.6±1.12	39.06±4.45
T3	18.14±0.78	33.48±1.73	11.6±1.60	40.67±4.30
T4	18.28±0.58	33.58±1.66	12.6±0.81	42.07±2.35
T5	19.7±0.75	34.44±2.31	11.4±1.31	38.94±1.379
T6	19.24±0.66	34.72±2.69	11.8±1.07	41.74±2.01

Table 5 represents the impact of vermi compost and soil mixture on the morphological characteristics of black gram. In general the root and shoot length of the plants grown increased gradually from the first day to 60 days and it was more than that of the control. Maximum growth was observed in the plants grown in T5. Nijhawan and Kanwar (1952) have observed similar results of increased root length than the control on application of earthworm compost to wheat. Maximum number of leaves was found on 60th day in the plants grown T4 and T6 of the worm worked compost and the leaves were minimum in numbers in the control pots. Nijhawan and Kanwar (1952) observed similar results in wheat. Leaf area index was also observed to be minimum in the control plants, which coincide with the work of Grappelli *et al.* (1985) in Salvia and Aster.

Yield characteristics of the cultivated black gram in the control and different treatments has been shown in Table-6. A maximum of 74.2 nodules were found in T6 treatment which is in agreement with the work of Kale *et al.*, (1992) who have reported significant increase in the colonization of the microbes in the experimental plots which received half the recommended dose of fertilizers and the vermicompost over control. The increased nutrient availability in the

Table-6. Impact of vermi compost and soil mixture on the yield of black gram

Treatment	No. of Nodules	No. of Flowers	No. of Pods	No. of Seeds	Wt. of Seeds (in cm.)
C	28.4±3.01	0.8±0.49	6.6±1.03	3.6±0.24	0.032±0.005
T1	41.2±6.04	2.2±0.58	10.2±1.56	5.2±0.37	0.037±0.006
T2	57.0±6.07	2.6±0.51	12.2±1.24	5.8±0.37	0.037±0.004
T3	64.8±2.82	2.2±0.58	11.8±0.66	5.8±0.37	0.049±0.002
T4	71.6±4.85	2.8±0.49	12.4±1.44	6.0±0.32	0.054±0.002
T5	68.8±1.56	2.4±0.93	13.6±1.66	6.0±0.32	0.054±0.002
T6	74.2.3.01±	2.8±0.97	14.4±1.03	6.2.0.37±	0.053±0.002

growth medium in T3 to T6 (Vermicompost, FYM and biofertilizers) may be the main reason for the maximum number of flowers in the present study which was also supported by Kang *et al.* (1994). The number of pods was found to be the maximum in T6 of worm worked compost and it was minimum in the control pots. Hopp and Slater (1949) quantified the improvement of crop yield due to earthworm addition. The maximum seed weight (0.054g) was recorded in T4 and T5 of *Lampito* compost. The reason for increased seed weight could be attributed that the leaf area index was also enhanced in the particular treatment, which could subsequently resulted in elevation of photosynthesis and accumulation in the grain.

Conclusion

The present study concludes that the use of vermi compost combined with FYM and biofertilizer will certainly enhance the yield of any crop besides enhancing the texture, colour, taste and even shelf life quality of the agricultural product.

References

Alexander, M. 1978: Introduction to Soil Microbiology. 2nd Edition. Wiley Eastern Limited, New Delhi. p. 467.

Allen, G. N. 1953: Experiment in soil bacteriology. Burgers Publ. Co. Minneopolis, Minn., USA. p. 127.

Balamurugan, V. 2002: Physico-chemical and microbial perspectives of vermicompost produced by *Eisenia fetida,* (Savigny), *Eudrilus eugeniae,* (Kinberg) and *Lampito mauritii,* (Kinberg) with a note on their fecundity and growth studies in ten different organic wastes. Ph.D. Thesis, Manonmaniam Sundaranar University, Tirunelveli, India.

Bhatnagar, R. K. and Palta, R. K. 1996: Earthworm - Vermiculture and Vermicomposting. Kalyani Publications, Ludhiana. p. 106.

Ceccanti, B. and Masciandaro, G. (1999). Vermicomposting of municipal and paper mill sludges. *Biocycle.* **6**: 71-72.

Gohil, M. B. 1995: Treatment of pulp and papermill waste. Case studies in pollution management in industries. (Ed.) Trivedy, R. K. Environmental Publications, Karad. p.16-20.

Graff, O. 1971: Stickstoff, phosphor und kalium in der Regenwurmlosung aif der Wiesenversuchsflache des Sollingprojektes. In: " IV Colloq. *Pedobiologia*". (Ed.) Agnika, D. Institut National des Recherches Agriculturelle, Publ. Paris. p. 71-77.

Grappelli, A., Tomati, V. and Galli, E. 1985: Earthworm casting in plant propagation. *Horti. Sci.* **20:** 874-876.

Guild, W.J. Mc, L. 1948: Effect of soil type on population. *Ann. Appl. Biol.* **35,** 181-192.

Haimi, J. and Huhta, V. 1986: Capacity of various organic residues to support adequate earthworm biomass for vermicomposting. *Biol. Fertil. Soils.* **2:** 23-27.

Hatanaka, K. Ishioka, Y. and Furuichi, E. 1983: Cultivation of *Eisenia fetida* using dairy waste sludge cake. In: Earthworm Ecology from Darwin to Vermiculture. (Ed.) Satchell, J. E., Chapman and Hall, London. p. 323-329.

HM Government. 1994: The Waste Management Licensing Regulations. Statutory Instrument. SI. No. 1056, HMSO, London.

Hopp, H. and Slater, C. S. (1949). The effect of earthworms on the productivity of agricultural soil. *J. Agric. Res.* **78:** 325-339.

Jambhekar, H. A. 1992: Use of earthworms as a potential source to decompose organic wastes. In: Proceedings of the National Seminar on Organic farming, Mahatma Phule Krishi Vidyapeeth, College of Agriculture, Pune. p. 52-53.

Kale, R. D., Mallesh, B. C., Bano, K. and Bagyaraj, D. J. 1992: Influence of vermicompost application on the available macro-nutrients and selected microbial populations in a paddy field. *Soil Biol. Biochem.* **24(12):** 1317-1320.

Kang, B. T., Akinnifesi, F. K. and Pleysier. 1994: Effect of agro forestry woody species on earthworm activity and physicochemical properties of worm cast. *Biol. Fertil. Soils.* **18:** 193-199.

Lee, K. E. 1985: Earthworms, Their Ecology and Relationship with Land Use. Academic Press, Sydney.

Nijhawan, S. D. and Kanwar, J. S. 1952: Physico-chemical properties of earthworm casting and their effect on the productivity of soil. *J. Indian Soc. Soil. Sci.* **22:** 357-372.

Raghuveer, S. and Sastry, C. A. 1995: Studies on biological treatment of pulp and paper mill waste waters. Waste treatment plants. Narosa public. House. New Delhi. p.401-405.

Tiwari, S. C. 1993: Effects of organic manure and NPK fertilization on earthworm activity in an Oxisol. *Biol. Fertil. Soils.* **16:** 293-295.

Umamaheswari, S., Selvakumar, S. and Vijayalakshmi, G. S. 2003a: Bioconversion of *Mangifera indica* litter by *Lampito mauritii* (Kinberg). Proc. on National Seminar on Global Strategies in Biological Sciences. Muthayammal College of Arts and Science, Rasipuram. p. 108.

Umamaheswari, Suveer, S. and Vijayalakshmi, G. S. 2003b: Vermicomposting of mango litter using *Perionyx* excavatus with a note on its physicochemical features. Proc on National Seminar on Global Strategies in Biological Sciences. Muthayammal College of Arts and Science, Rasipuram. p. 108.

Chapter 12

Bio-management of Banana Packing Waste using an Epigeic Earthworm, *Eisenia fetida*

Satyendra M. Singh, Geeta R. Gangwar & Om Prakash
Vermiculture and Environmental Research Laboratory, Department of Animal Science, Mahatma Jyotiba Phule Rohilkhand University, Bareilly-243 006 (U.P).

Generation of solid waste is one of the serious environmental, social and economic issues throughout the world. The quantity of solid wastes generated is increasing rapidly with growing economic activities and the production and use of consumer items. In Asia-pacific region, the total amount of waste generated each year is about 2.6 billion tons. Senapati (1993) has reported that the annual production of organic wastes in the country was about 3000 million tons. India generates 377 million tons crop residues as plant wastes (Reddy and Rao, 1998). During transportation of banana, their leaves are used as packing material. There are one of such wastes accumulated in huge amounts everyday in all the fruit markets of the country. It has been estimated that about 166 tons of such waste is thrown out per day on roadsides from wholesale fruit market of Bareilly city of Uttar Pradesh state alone. It is decompose at a very slow pace on different sites creating environmental hazards and is considered as one of the absolute waste. In present studies this plant waste has been taken along with different proportions of cattle dung to see its rate of transformation into vermi compost using an epigeic earthworm, *Eisenia fetida*. Growth of earthworms in terms of increase in their number and biomass and rate of cocoon production were also investigated. Assessment of nutritional status of waste transformed vermi compost on *Phaseolus aureus* - one of the common leguminous plants was carried out to see the level of its utility on the plant growth.

Vermi compost preparation

Banana leaves, collected from wholesale fruit market of Bareilly, were chopped in to small pieces of ± 3 cm. sizes, shade dried in Animal House of the Department. Leaf pieces were mixed with cattle dung in 1:3,1:1 and 3:1 ratios separately. Six tubs (43 x 32 x 14 cm size) of each ratio were prepared, each having 4 kg mixed medium and subjected to pre-decomposition process for a fortnight. Nearly 40% moisture level of each media was maintained by sprinkling water.

In first three tubs of each mixed waste material, 10 g mature (clitellate) epigeic earthworms, *Eisenia fetida* were introduced and remaining three of each medium were kept as control - devoid of earthworms. Experiment was carried out at 20-25^{oC} room temperature till the transformation of waste into vermi compost. Cocoons and earthworms were sorted out from the prepared vermicompost for their number and weight, respectively. Physico-chemical parameters of every waste media and vermicompost were analyzed using standard known methods before and after experimentation.

Effect of 10 % aqueous extract of different waste media transformed vermicomposts on the germination of *Phaseolus aureus* was assessed. A filter paper soaked with the extract of vermicompost of 1:3 waste medium was kept in a petridisc (size 6"diameter) and seven seeds of *P.aureus* were put on it. Another filter paper soaked with the same extract was kept on the seeds. A ± 2 cm thick cotton pad was put on the covered filter paper. Moisture was maintained by wetting the cotton pad with the same extract as and when required. Similar experiments were carried out for all other vermicomposts and plane soil extract. The size of plumules and radicals was measured after 4 days and the data were compared with the germinated seeds in the extract of normal soil.

Impact of different concentrations of vermicompost (1:3, 1:1 and 3:1 as vermi compost : soil) on the shoot length of moong (*P.aureus*) was assessed by using earthen pots (surface diameter ±11.6 cm), in triplicate. Three healthy seeds of *P.aureus* were sown in each pot at the depth of ±1 cm from the surface. Moisture content was maintained by sprinkling appropriate amount of water alternatively. Length of shoot was measured after 10, 20, 30 and 40 days and the data were compared with the plant grown in normal soil.

Experimental findings

i) Number and Biomass of Earthworms

Number and biomass of earthworms, *E. fetida,* before and after the bioprocessing of banana packing waste into vermicompost have been shown in (Table-1). It showed that there was a loss in number of clitellates although their total weight was nearly constant indicating that those survived, fed upon the waste media and reproduce well. Rate of reproduction of *E. fetida* was 10.2 times faster in 1:3 waste medium than that of 3:1 and 1.7 times only than that of 1:1 medium. Rate of production rate of cocoons was recorded 3.2 times more in 1:3 mediums than that of 1:1 and 3:1 media. It is also noted that the increase in number of total earthworm population was 8 times in 1:3 than 3:1 and 4.5 times in 1:1 medium than that of 3:1. This increase was 1.6 times in 1:3 with respect to 1:1. Increase in biomass of earthworm was 3.5 times in 1:3 than that of 3:1 and 2.3 times with that of 3:1. It was only 1.5 times in 1:3 with respect to 1:1. It is further noticed that the hatching rate of cocoons was much faster in 1:3 medium producing the maximum number of juveniles than that of other two media. (Fig.1). Slowest rate of transformation of waste and reproduction and growth of worms in 3:1 medium may be due to lower level of moisture content and higher pH (Table-2) than that of other two media. Reinecke and Venter (1985) also reported that lowering of the growth rate due to low moisture conditions would also retard sexual development. Dresser and Mckee (1981) found that moisture content between 50 and 80% is the most appropriate for vermicomposting process. *E.fetida* can survive in the moisture range between 50 and 90% and grows more rapidly between 80-90% (Edwards *et al.,* 1985). Edwards (1988) has reported that worm's growth and reproduction becomes faster if the pH of the medium is in the range of 5 to 9 and temperature between 15 to 20°C. Present findings showed similar results on the growth and reproduction of the worms in the plant waste.

ii) Factors affecting conversion rate of waste by earthworms

The rate of conversion of waste by earthworms into vermicompost was noticed faster in 1:3 medium. It might be due to presence of 2.1 times more percent organic matter in it than that of 3:1.Significant decrement in % moisture content, organic carbon and pH, C: N ratio during the process of vermicomposting and increase in percent level of macronutrients (N and P) in all media showed the role of intestinal

Table1: Showing number and biomass of earthworm, *E. fetida* before and after vermicomposting of banana packing leaves (*Musa paradisiaca*) waste (SE=± 3)

Ratio of the medium (waste:cattle dung)	Earthworm population (before vermin composting)		Earthworm population (after vermin composting) in three age groups						Rate of reproduction (young worm-1 week-1)	Number of cocoons
	Clitellates		Juveniles		Non-clitellates		Clitellates			
	Number (±)	Weight (±g)	Number (±)	Weight (±g)	Number (±)	Weight (±g)	Number (±)	Weight (±g)		
1:3	38±0.66	10	296±12.0	15.35±0.77	76±6.8	10.62±0.60	30±1.3	10.07±0.66	1.12±0.04	54±5.5
1:1	37±0.61	10	60±6.3	2.35±0.71	168±7.0	14.73±0.11	33±1.2	9.72±0.33	0.65±0.03	18±3.7
3:1	43±0.66	10	27±5.0	1.05±.0.24	23±3.0	3.83±0.45	38±1.1	12.54±0.21	0.11±0.01	17±4.2

Table-2: Physico-chemical parameters of banana packing leaf waste (*Musa paradisiaca*) before and after vermicomposting

Initial parameters (before inoculating worms) into waste: dung media				←Ratios→ of banana leaf waste: cattle dung used	Final parameters (after vermicomposting)						
					Experimental			Control (without worms)			Pure waste (without Dung and worms)
1:3	1:1	3:1	Pure waste (without dung and worms)	↓Parameters measured	1:3	1:1	3:1	1:3	1:1	3:1	
82.52	83.41	82.53	67.82	% Moisture content	63.46	57.87	60.0	70.05	68.08	67.21	65.54
9.95	9.96	10.16	10.26	pH	9.31	9.46	9.51	9.36	9.58	9.62	9.68
26.38	14.08	12.63	12.78	% Organic matter	19.80	12.82	11.35	20.35	12.72	11.83	10.12
15.33	8.18	7.34	7.43	% Organic carbon	11.50	7.45	6.59	11.83	7.39	6.87	5.93
2.3	1.1	1.7	1.0	% Nitrogen	3.2	3.1	2.5	2.9	1.3	1.8	1.7
6.47	7.11	4.14	7.0	C: N	3.71	2.28	2.63	4.07	5.55	3.69	3.44
0.20	0.15	0.10	0.10	Phosphorus (%)	0.30	0.25	0.15	0.20	0.15	0.10	0.10
1.080	1.258	1.216	1.092	Zn (ppm)	1.793	1.880	1.482	1.209	1.193	0.996	1.006
20.17	28.16	25.74	12.526	Fe (ppm)	24.98	24.27	17.46	24.52	20.25	15.70	9.86
0.242	0.219	0.185	0.128	Cu (ppm)	0.333	0.333	0.214	0.195	0.231	0.151	0.139
14.97	14.54	12.33	10.23	Mn (ppm)	11.79	15.32	13.32	10.52	13.49	10.75	14.64

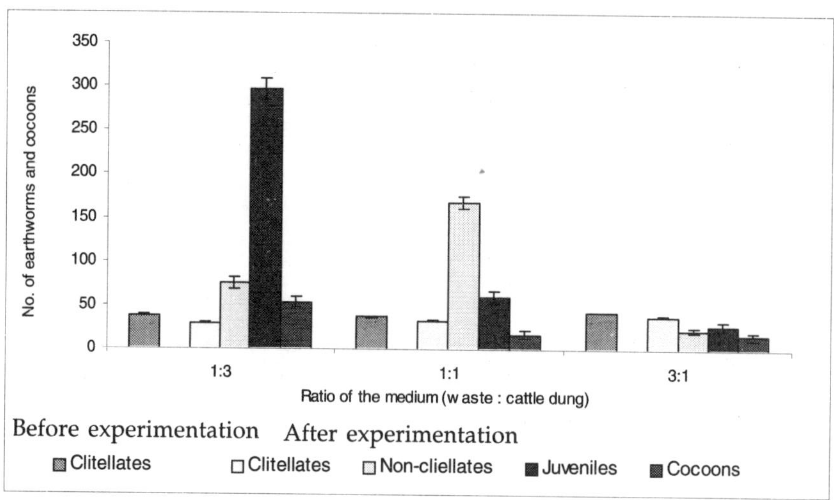

Fig. 1 : Showing the number of earthworm, *E. fetida* before and after vermicomposting of banana leaves (*Musa paradisiaca*) waste (SE=±3)

micro-organisms of earthworms and secretions from their body during bioprocessing of the plant leaf waste.

Analysis of data showed that the maximum decrement (24.98%) in % organic matter and carbon was in the medium of 1:3; while it was the minimum (8.94%) in 1:1 ratio of experimental medium. However, it was the maximum (22.85%) in the 1:3 ratio and the minimum (6.33%) in 3:1 ratio of control medium. This decrement was 20.10% only in pure banana leaf waste medium after the completion of experiment. These findings indicate that the rate of decrement of the organic matter and carbon was 2.13% and 2.61% more in 1:3 and 1:1 experimental media, respectively, than that of their respective controls.

It was also noticed that the per cent of nitrogen was increased by 2.8 and 1.3 times more in 1:1 and 1: 3 ratio of experimental media; while only 1.1 and 1.0 times in 1:1 and 1:3 ratio of control media. In pure waste medium, such increment was 1.7 times after the completion of the experiment with respect to its initial value. Edward and Bohlen (1996) has also reported that earthworms consume large amounts of plant organic matter that contain considerable quantities of nitrogen and much of the nitrogen that they assimilate into their own tissues is returned to the soil in their excretions. He further stated that the earthworms can alter the C: N ratio of material that process through their digestive tract although they have relatively low assimilation efficiencies for both carbon and nitrogen.

In present findings, C: N ratio was decreased by 67.47 % in 1:1and 36.47% in 3:1 ratio of experimental media; while in the control, it was more decreased (37.09%) in 1:3 and less decreased (10.86%) in 3:1 ratio. Higher level of decrement of C: N ratios in all experimental media showed the level of increase of nitrogen content in the waste transformed into vermicompost than that of their respective controls.

Phosphorus - one of the important macronutrient for energy metabolism of plants was increased by 0.66% in 1:1 and 0.50% in 1:3 and 3:1 in both experimental media. The level of micronutrients was increased significantly in all the experimental media than that of their respective controls.

10% aqueous extract of vermicompost promoted the germination rate of radicle and plumule of seeds of *Phaseolus aureus* (Fig.2). Rate of germination was the maximum in the extract prepared from the vermicompost of 1:3 ratio medium; while the minimum in 3:1. Growth rate of radicle was 1.5 times more in the aqueous extract of vermicompost prepared from 1:3 ratio than that of 3:1; while it was only 1.4 times in case of plumule of the seeds. It may also be observed

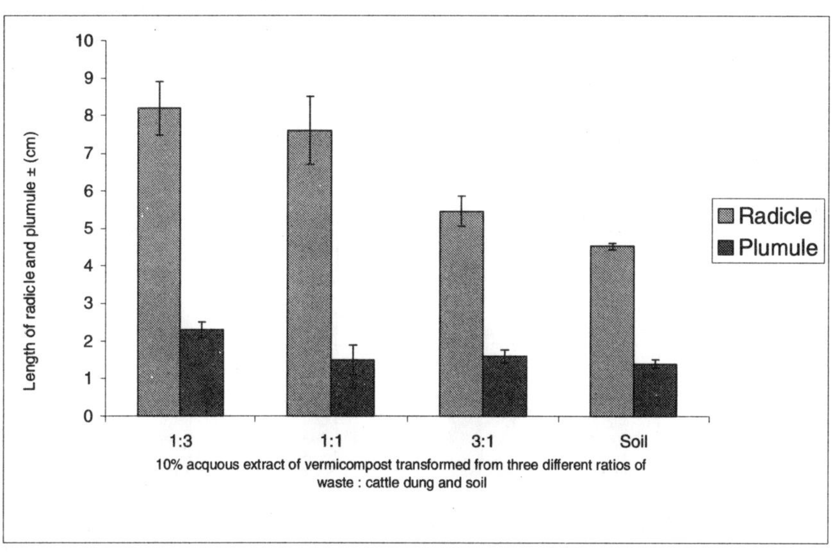

Fig. 2 : Germination of moong seed (*Phaseolus aureus*) in 10% aqueous extract of vermicompost of banana leaves (*M. paradisiaca*) in terms of length of radicle and plumule) (SE=±3)

that the rate of germination of seeds of *P.aureus* was slower than that of all types of aqueous extracts of vermicompost if only soil was used as one of the mediums.

Effect of vermicomposts transformed from banana leaf waste media with different proportions of cattle dung on the shoot length of *P. aureus* has been shown in Fig.3. It was recorded that the medium having vermicompost (which was transformed from 1:3 ratio of waste and cattle dung) and soil in the ratio of 1:3 was the best for the shoot length of the experimental plants than that of the others. Different ratios of vermicompost and soils used to see their impact on shoot length of the plants showed slight variations that would be due to accumulation of higher level of available nutrients during the process of vermicomposting. Edwards (1995) has also mentioned that the important plant nutrients in the organic material particularly nitrogen, phosphorus, potassium and calcium are released and converted through microbial action into forms that are much more soluble and available to plants that those in the parent compounds.

Conclusion

It was recorded that 1:3 waste medium is better for the reproduction of the worms and production of cocoons, their hatching and the rate of transformation of waste into vermicompost faster than that of other

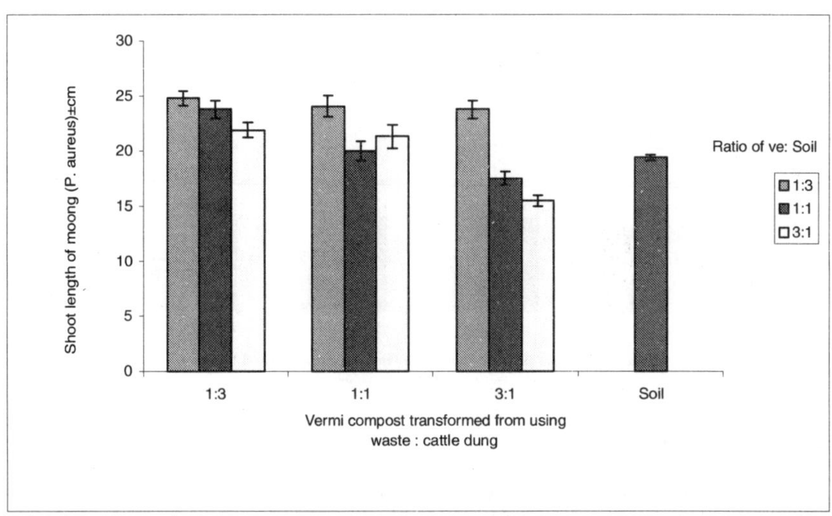

Fig. 3 : Showing shoot length of moong (*P. aureus*) in banana leaf waste transformed vermicompost after 40 days of experimentation (SE=±3)

two experimental media used. However, increase in the amount of phosphorus was 0.66% in 1:1 medium and only 0.50% in other two media.

10% aqueous extract of vermicompost prepared from 1:3 ratio waste medium has better impact on the germination of seeds of *P. aureus* than that of the extracts of other two varieties of vermicompost transformed from 1:1 and 3:1 waste media. Germination rate was slowest when aqueous extract of plane soil was applied on the seeds.

Impact of transformed vermicompost on the shoot length of *P. aureus* with different ratios of soil showed that the vermicompost transformed from 1:3 waste media was suited the best if it is applied with the soil in the ratio of 1:3 than the vermicompost if mixed with soil in ratio of 1:1 and 3:1.

References

Dresser C. and McKee I., 1981: Compendium on solid waste management by vermi composting. Prepared for the Municipal Environmental Research Laboratory office and development, Environmental Protection Agency, Cincinnati.

Reinecke A.J. and Venter J.M., 1985: Moisture preferences, growth and reproduction of the compost worm, *Eisenia fetida* (Oligochaeta). Biology and Fertility of soils 3, p. 135-141.

Edwards, C.A., 1995: Historical overview of vermicomposting. Biocycle. p. 56-58.

Edwards, C.A. 1988: Breakdown of animal, vegetable and industrial organic wastes by earthworms. Agric.Ecosyst.Environ.24, p. 21-31.

Edwards, C.A. and Bohlen, P.J., 1996: Biology and Ecology of Earthworms. 3rd edi. Chapman and Hall, London.

Edwards, C.A., Burrows, I., Flectcher K.E. and Jones B.A., 1985: The use of earthworms for composting form wastes in composting of Agricultural and other wastes (J.K.R.Grasser, Eds), p.229-241.Elsevier Applied Science, Oxford.

Reddy, D.D. and Rao, A.D., 1998: Crop residues- A renewable resource to sustain soil health and crop production. Farmer and Parliament. 35 (8): p.38-39.

Senapati, B.K.1993: Vermitechnology in India. Earthworm resources and vermiculture. Ed. A.K. Ghosh. Zoological Survey of India, Calcutta. p. 109-111.

Chapter **13**

Vermicomposting of Cereal and Vegetable Crop Wastes using Earthworm, *Perionyx excavatus*

A.K.Sannigrahi
Proof & Experimental Establishment, Chandipur, Balasore - 756 025 (Orissa), India.

Adoption of mechanized farming by the farmers facilitates production of a sizable quantity of crop residues as leftover in the field after harvesting. Zadrazil (1978) reported that more than half of total produce from land remained unused as crop wastes in the form of straw, leaves, stems, roots, etc. About 377 million tons of crop residues are available every year in India (Reddy and Rao (1998). Sidhu (1998) estimated that out of 28.8 million tons of wheat and paddy straw produced annually in Punjab, the farmers in the field burned about 17 million tons. Sugarcane trash is either burnt in the field or used as fuel for cooking, as it is hard to decompose. Vegetable crop wastes are either discarded in the field or dumped on roadside or municipal dustbin as leftover of vegetable market. On an average, 25 % of total nitrogen, 50 % of total phosphorus, 75 % of total potassium and 50 % of total sulfur taken by the crops are retained in crop residues. Sharma and Mishra (2001) reported that complete burning of paddy and wheat straw resulted about 100,100,20.1,19.8 and 75 % loss of carbon, nitrogen, phosphorus, potassium and sulfur, respectively .The practice of burning of crop wastes causes a great loss of valuable plant nutrients from soil resulting loss of its fertility as well as productivity. Burning of one-kilogram crop waste also produces about 20-114 g carbon mono oxide and 2.1-11.4 g total suspended particles as smoke resulting environmental pollution (Chinnamani, 1992). Municipal authorities often face problems for disposal of vegetable crop wastes generated in the vegetable market (Sabji mandi), especially the rotten onion, tomato, potato etc.due to their foul smell. Being organic in nature, crop wastes can be converted to compost quickly following vermicomposting technique (Bhawalkar, 1995;Sannigrahi and Chakrabortty, 2000). Sannigrahi (2005) showed that both *Perionyx excavatus*, the local breed of earthworm and *Eisenia fetida*, the commercial breed of earthworm were equally efficient in vermicoposting of crop wastes and nutrient enrichment of compost in humid sub tropical climate of North East India. Application of compost in the field improves soil fertility and increases crop production (Edwards and Burrows, 1988; Parthasarthi and Ranganathan, 2002; Sannigrahi, 2002). Burning or non-composting of crop wastes causes loss of valuable plant nutrients, the great economical loss to nation. Detailed information about the quality of vermicompost prepared from different crop wastes is very limited. Keeping these in view, vermicomposting of different cereal and vegetable crop wastes was carried out using earthworm, *Perionyx excavatus*, to estimate time required for vermicomposting and analyze the quality of vermicompost prepared from different crop wastes.

Collection of plant wastes

Three types of cereal straw, four types of other crop wastes, twelve types of vegetable crop wastes and three types of waste vegetables were collected for the experiment. These were maize plants (*Zea mays*), paddy straw (*Oryza sativa*), wheat straw (*Triticum durum*), ground nut plants (*Arachis hypogea*), lemon grass (*Cymbopogon caesius*), mustard plants (*Brassica campestris var.dichotoma*), sugarcane bagasse (*Saccharum officinarum*), broccoli leaves (*Brassica oleracea var.botrytis*), brussel sprout leaves (*Brassica oleracea var. gemmifera*), cabbage leaves (*Brassica oleracea var. capitata*), cauliflower leaves (*Brassica oleracea var. botrytis*), French bean plants (*Phaseolus vulgaris*), knolkhol leaves (*Brassica oleracea var. gongylodes*), lady's finger plants(*Abelmoschus esculentus),* lai leaves(*Brassica juncia var. cuneifolia*), onion plants (*Allium cepa*), potato plants (*Solanum tuberosum*), radish leaves (*Raphanus sativus*), tomato plants (*Lycopersicon lycopersicum*), lady's finger fruit peel, rotten onions and tomatoes.

Vermicomposting of plant wastes

It was carried out using "Two stage vermicomposting technology"(Sannigrahi and Sannigrahi, 2006). Each 20 kg crop waste was mixed with slurry of 40 kg cow dung and heap like rectangular beds of size 2 feet X 1.5 feet were made over cemented floor under the shade, in triplicate, following randomized block design. All beds were covered with black polythene sheets to facilitate rising in the initial temperature of the beds, quick decomposition of waste materials and avoiding breeding of flies and other insects. After 15 days, the waste mixture of bed was thoroughly mixed to avoid any dry un-decomposed portion, heaps were made again and 500 locally available earthworms, *Perionyx excavatus*, were released on the top of each bed and all beds were covered with moist gunny bags for facilitating proper moisture and aeration. Regular monitoring was done to maintain 70 to 80 % moisture in the beds by spraying water and to protect earthworms from predators like frogs, birds, mongoose, red ants etc. Earthworms deposited vermicompost on the top as excreta. When the composting material became black coloured loosely granular in shape up to lower layer as observed by physical verification once in a week, whole material of the bed was spread in the shade for bringing down the moisture at about 40 % by air-drying. Prepared vermicompost of each of the bed was sieved separately using 2 mm mesh and samples were analyzed in the laboratory using following standard procedures (Tandon, 1993).

Total nitrogen was estimated by Kjeldahl digestion and Distillation method, pH of samples was determined by taking compost: water mixture in the ratio of 1:2 using pH meter. Acid digestion of samples was carried out using concentrated HNO_3 and $HClO_4$ mixed in 9:4 ratios. Digested extracts were used to determine total potassium, sodium and calcium using Flame photometer; while phosphorus and sulfur by Spectro photometer. Results were calculated and analyzed statistically.

Experimental findings

Tables 1 and 2 represent the analytical data of different vermicompost prepared from seven cereal and other crop wastes and fifteen vegetable crop wastes, respectively. Vegetable crop wastes, being soft in nature, took lesser time (1 to 2 months) for complete vermicomposting as compared to hard natured cereal and other crop wastes (3 to 8.5 months). Sugarcane bagasse took about 8.5 months for vermicomposting. The pH of vermicompost prepared from sugarcane bagasse was found slightly acidic; while those of lemon grass, onion plants, potato plants and radish leaves were found slightly alkaline. This was mainly due to presence of more sodium, potassium and calcium ions in the vermicompost as compared to others. The pH of other vermicompost varied between 6.87 to 7.90 indicating their neutral nature. The pH of vermicompost was slightly higher than the respective raw waste materials due to mixing of amorphous calcium carbonate particles coated with mucus discharged from the calciferous glands of earthworms during vermicomposting (Senapati, 1993). At the initial stage earthworms use to crawl away from radish wastes due to its pungent smell. Total nitrogen, phosphorus, potassium, sodium, calcium and sulfur contents of these vermicompost varied from 0.79 to 2.03 %, 0.18 to 0.96 %, 1.20 to 3.12 %, 1.11 to 3.22 %, 0.72 to 3.75 % and 0.37 to 1.91 %, respectively. Chemical analysis confirmed the preparation of good quality vermicompost from different types of crop residues. The amount of plant nutrients in vermicomposts varies with the variation of crop waste materials. Even different vegetable crop wastes can be beneficially utilized after transforming into vermicompost.

Table 1. Chemical analysis of vermicompost prepared from differnet crop wastes

Crop wastes (cereals and others	Composting Period (month)	pH (1:2)	Total N (%)	Total P (%)	Total K (%)	Total Na (%)	Total Ca (%)	Total S (%)
Maize plants	3.5	7.83	0.79	0.45	2.08	1.74	0.87	0.68
Paddy straw	3.0	7.87	1.45	0.71	2.56	2.39	1.11	0.99
Wheat straw	3.0	7.87	0.99	0.49	2.60	2.38	1.37	1.07
Ground nut plants	3.0	7.33	1.21	0.81	2.02	1.71	1.16	1.91
Lemon grass	5.0	8.33	1.25	0.76	1.81	1.67	1.94	1.38
Mustard plants	3.0	7.30	1.62	0.84	1.52	1.06	1.71	0.80
Sugarcane bagasse	8.5	6.30	1.20	0.75	1.49	1.49	1.33	0.92
CD at 5%		0.31	0.10	0.07	0.09	0.08	0.08	0.11

Table 2. Chemical analysis of vermicompost prepared from differnet crop wastes

Crop wastes (cereals and others	Composting Period (month)	pH (1:2)	Total N (%)	Total P (%)	Total K (%)	Total Na (%)	Total Ca (%)	Total S (%)
Broccoli leaves	2.0	7.30	1.66	0.77	2.10	2.18	1.74	1.23
Brussel sprout leaves	2.0	7.13	1.55	0.82	2.10	2.08	1.68	1.29
Cabbage leaves	1.0	7.03	1.47	0.69	1.79	1.98	1.55	1.01
Cauliflower leaves	1.5	7.20	1.22	0.73	1.80	1.92	1.76	1.08

French bean plants	2.0	7.13	1.42	0.87	1.94	1.33	0.78	1.23
Knolkhol leaves	1.0	6.87	1.56	0.55	1.50	1.41	2.06	1.17
Lady's finger plants	3.5	7.87	0.91	0.46	1.71	1.68	1.09	0.89
Lai leaves	2.0	6.87	1.27	0.61	1.37	1.44	1.29	0.65
Onion plants	1.0	8.13	1.66	0.76	1.67	1.71	1.05	1.01
Potato plants	1.5	8.73	1.24	0.71	3.12	3.22	2.99	0.91
Radish leaves	2.0	8.77	1.29	0.80	2.62	1.88	1.20	1.44
Tomato plants	1.5	7.37	2.03	0.86	2.62	2.39	2.33	1.21
Lady's finger fruit peel	2.5	7.90	1.04	0.18	1.74	2.05	3.75	0.58
Onion (rotton)	1.5	7.47	0.91	0.43	1.20	1.11	0.72	0.37
Tomato fruit (rotton)	1.5	7.43	1.98	0.96	2.53	2.45	1.25	0.72
CD at 5%		0.28	0.12	0.06	0.12	0.11	0.09	0.10

Conclusion

It may be concluded that the crop wastes are not the waste but essential raw materials for the production of vermicompost. Being a marketable product, vermicompost is sold at the rate of 4 to 5 rupees/ kg in eastern India. There is a huge demand for vermicompost in the country and both Central and State Governments are encouraging sustainable organic farming systems. Even tea gardens in Northeast are gradually switching over to organic tea production for better export opportunity. 'Two stage vermicomposting technology' is not only easy to follow by all peoples irrespective of their age, sex, caste or educational level but also helps in commercial production of solid vermicompost for soil application and liquid vermi-T for spraying over standing crops. People can go for vermicomposting as side business without disturbing their main job. This technique also offers more scope for employment generation and extra income even at village level. Success will be achieved if awareness about reuse of crop wastes increases among common people, their burning or throwing should be stopped and vermicomposting could be started as a cooperative business similar to that of "Amul"production in Gujarat.

References

Bhawalkar, V.S.1995: Sustainable agriculture through Vermiculture. Indian farmers Times.13: 13-14.

Chinnamani, S.1992: Agro forestry for fuel resource management. Proc.XII.National symposium on Resource Management for sustained crop production. Rajasthan Agricultural University, Bikaner, 25-28 February p.351-356.

Edwards, C.A. and Burrows, I.1988: The potential of earthworm composts as plant growth media. In: Earthworms in Environmental and Waste Management (C.A. Edwards and E.F.Bneuhauser, Eds), SPB Acad. Publ., The Netherlands, p.211-220.

Parthasarathi, K.and Ranganathan, L.S.2002: Supplementation of press mud vermicasts with NPK enhances growth and yield in leguminous crops (*Vigna mungo* and *Arachis hypogea*), J.Curr.Sci.2: 35-41.

Reddy, D. D. and Rao, A.D.1998: Crop residues-a renewable resource to sustain soil health and crop production. Farmers and Parliament.35 (8): 38-39.

Sannigrahi, A.K.2002: Vermicomposting of biodegradable house refuses for reducing municipality solid waste load. In: Solid Waste Management -Current status and strategies for the future (R.K. Somashekar and M. A. R. Iyengar, Eds), Allied Publ. Pvt. Ltd., New Delhi, p.107-111.

Sannigrahi, A.K.2005: Efficiency of *Perionyx excavatus* in vermicomposting of thatch grass (*Imperata cylindrica*) in comparison to *Eisenia fetida.* Assam. J. Indian Soc.Soil Sci.53 (2): 237-239.

Sannigrahi, A.K.and Chakrabortty, S.2000:Impact of composting techniques on the availability of NPK and microbial population in composts. Environ. & Ecol.18 (4): 888-890.

Sannigrahi, A.K.and Sannigrahi, Debanjan, 2006:Two stage composting technique for rapid and beneficial utilization of firm wastes. Indian Farmers' Digest 39(11): 17-21.

Senapati, B.K.1993: Earthworm gut contents and its significance. In: Earthworm Resources and Vermiculture (A.K. Ghose, Ed), Zoological Survey of India, Calcutta, p.97-99.

Sharma, P.K. and Mishra, B. 2001:Effect of burning rice and wheat crop residues in loss of NPK and S from soil and changes in the nutrient availability.J.Indian Soc.Soil Sci.149 (3): 425-429.

Sidhu, B.S.1998: Sustainability implications of burning rice wheat straw in Punjab. Economic and Political Weekly, Sept.26.

Tandon, H.L.S.1993: Method of analysis of soils, plants, water and fertilizers. Development and Consultation Organization, New Delhi, p.13-35.

Zadrazil, F.1978: Cultivation of pleurotas. In: The Biology and Cultivation of edible Mushroom (S.T. Chang and W. A. Hayes, Ed), Academic Press, New York, p.521-558.

Chapter **14**

A Primary Study on Vermicomposting of Acidic Substrate Through Microbial Management

Souri Roy*, G.N.Chattopadhyay and Sujit Mal
Soil Testing Laboratory, Institute of Agriculture, Visva-Bharati, Sriniketan-731236, West Bengal
**Cyber Research and Training Institute, Burdwan-713101, West Bengal*

The importance of earthworms in maintaining productivity of soils has been known since long (Darwin, 1837). Beneficial effects of these earthworms are improving structure, aeration, nutrient status and some other properties of the soils and, thereby, the growth of crops has been well documented (Edwards and Lofty, 1972;Ismail, 1997). However, the concept of recycling different kinds of organic waste materials as good quality compost through the help of gut micro-organisms of epigeic earthworms has emerged in recent past (Senapati, 1993; Kale, 1998). The intestine of these earthworms harbour high concentrations of different kinds of micro-organisms (Wallwork, 1984;Satchell, 1967). The food materials ingested by these earthworms are thus subjected to more intense microbiological activity in their alimentary canals and also after their excretion as vermicast to form nutrient rich vermicompost (Chattopadhyay, 2005). Although this biotechnology involves recycling of wide ranges of organic wastes with the help of large numbers of earthworm gut microorganisms, yet the efficiency of this composting process has been observed to the rather limited during the use of any acidic waste material as the substrate. This behaviour may be attributed to sensitivity of the earthworms and also most of the decomposer microorganisms to acidity of the substrates thus restricting their activities. The problem may be overcome by either reducing the acidity of the organic waste through use of any ameliorant or by inoculation of some acidity tolerant microorganisms in the decomposition process or by combining both the techniques. However, information on this aspect with regard to vermicomposting is very meager. In the present investigation, therefore, an attempt has been made to assess the possibility of increasing the efficiency of vermicomposting biotechnology on an acidic substrate by using different combinations of lime, for increasing the pH of the substrate, and *Trichoderma viridae*, an acidity tolerant fungus. It is hoped that the results of this primary study will generate some useful information leading to development of an effective package for vermicomposting of acidic organic wastes.

Processing of organic substrate

The study was carried out in yard using earthen containers of 1kg capacity. Patharkuchi (*Kolanchoe pinnata*), a commonly observed wild herb with acidic vegetative parts was used as the organic substrate for the study. The chopped plant parts were mixed with cow dung at 3:1 ratio and taken in different earthen pots @ 1kg per pot. Initial pH of the organic waste mixture was 5.5. The mixed organic materials were incubated under moist condition after which the treatments were added to these materials. Five treatments viz. control TV, $CaCO_3$, $CaCO_3$ + TV (1/2), $CaCO_3$ (1/2), $CaCO_3$ (1/2) + TV (1/2) were used for the study. The treatment with full dose of $CaCO_3$ included use of 1500 mg $CaCO_3$ per kg of organic waste while that of *Trichoderma viridae* (TV) consisted of inoculation with 10 x 10^6 nos. of TV per kg waste. The treatments marked half (½) indicated use of 50% of the respective doses. After a week of these treatments, epigeic earthworm, *Eisenia fetida* was released to each pot at 10 nos. of worms per kg of organic waste. The vermicomposting with different worms was carried out for 45 days. However, during the initial phase, the earthworms did not survive well in the treatments without lime and had to be replenished periodically for about 15 days. At the end of the incubation, the composted materials were analyzed for pH, easily mineralisable nitrogen (Subbiah and Asija, 1954), easily oxidisable organic carbon (Walkley and Black, 1934) and microbial biomass carbon (Franzluebbers, 1999) for assessing the quality of the composts under different treatments.

Experimental findings

Vermicomposting is carried out by the intestinal microorganisms of epigeic earthworms (Tripathi, 2005). However, in addition to susceptibility of the earthworms, majority of these gut micro-flora viz., bacteria and actinomycetes also cannot survive well in acidic substrates (Subba Rao, 1999). Therefore, the rate of vermicomposting declines with the use of such materials as the decomposing substrate. In the present investigation, the initial pH of the organic material was 5.5 that inhibited the gut microorganisms in the substrate and thus slowed down the vermicomposting process. This behaviour was reflected in occurrence of highest amount of easily oxidisable organic carbon content and also the lowest value of easily mineralisable nitrogen of the final product in the control treatment (Table - 1) indicating that the rate of decomposition was slowest under this treatment.

Table-1 Changes in some chemical properties of vermicomposted substrates under different treatments

Treatment	pH	Easily mineralisable N (mg kg^{-1})	Easily oxidisable organic C (%)
Control	6.0	3523.4	25.3
TV	6.5	4600.9	19.5
$CaCo_3$	7.2	4388.4	18.5
$CaCo_3$ + TV ($^1/_2$)	7.1	6578.8	16.9
$CaCo_3$ ($^1/_2$)	6.9	4526.2	18.2
$CaCo_3$ ($^1/_2$) + TV ($^1/_2$)	7.0	5726.8	17.8

TV = *Trichoderma viridae* @ 10^6 nos. kg^{-1}.

$CaCO_3$ = Lime as $CaCO_3$ @ 1500 mg kg^{-1}.

TV ($^1/_2$), $CaCO_3$ ($^1/_2$) = 50% of respective dose.

The situation improved with the treatment using *Trichoderma viridae* (TV) resulting in an increment of easily mineralisable nitrogen status to 4600.9 mg kg^{-1} and reduction of easily oxidisable organic carbon value to 19.5%, which showed that the rate of vermicomposting proceeded more favourably under this TV treatment. While most of the earthworm gut micro flora is susceptible to acidic condition, TV fungi were tolerant to this environment and carried out the decomposition of the organic materials more effectively. However, the earthworms were not very comfortable in this treatment especially during the initial phase and had to be replenished partly during first 15 days of incubation. Use of two doses of lime also tended to increase the rate of decomposition of the acidic substrate probably owing to occurrence of more favorable environment to the earthworms and also other micro-organisms under the near mental pH ranges of the organic wastes. However, the treatments using combination of lime and TV appeared to be the most effective in the composting process. This was attributed to increased comfort level of the earthworms in the near neutral pH values under the treatments with lime and also the increased activity of different kinds of gut microbes including TV in decomposing the substrate.

These observations have been supported by the values of microbial biomass carbon (MBC) under different treatments that can provide gross idea about the quantum of living components in the organic matters under a particular system (Bailey *et al.*, 2002). In the present study, the control treatment showed the lowest MBC value obviously owing to restricted microbiological activity under the acidic environment. Inclusion of TV in the composting system resulted in considerable increment in MBC value due to survival of this acid tolerating fungi in the substrate. However, it was interesting to observe that although no external source of microbe was introduced to the composting system, yet the MBC values under the two line treatments increased considerably. This was probably due to generation of some micro-organisms under the favourable environment of lime treated system. It was interesting to observe that the MBC values remained under the highest ranges in both the $CaCO_3$ + TV treatments and these treatments resulted in more rapid rate of decomposition, as was reflected in high mineralisable nitrogen values and the magnitude of reduction in easily oxidisable organic carbon status of the composts. This behaviour could be attributed to increased concentrations of TV in these treatments and their activity in degrading the substrate. The efficiency of *Trichoderma* in decomposing more resistant components of organic matters has been reported by Gaur and Singh (1995).

Conclusion

This primary study showed that the decomposition of acidic organic wastes under vermicomposting could be improved considerably through use of lime and adoption of microbial management. However, more detailed investigations need to be carried out with regard to behaviour of earthworms in such substrates and also the roles of the

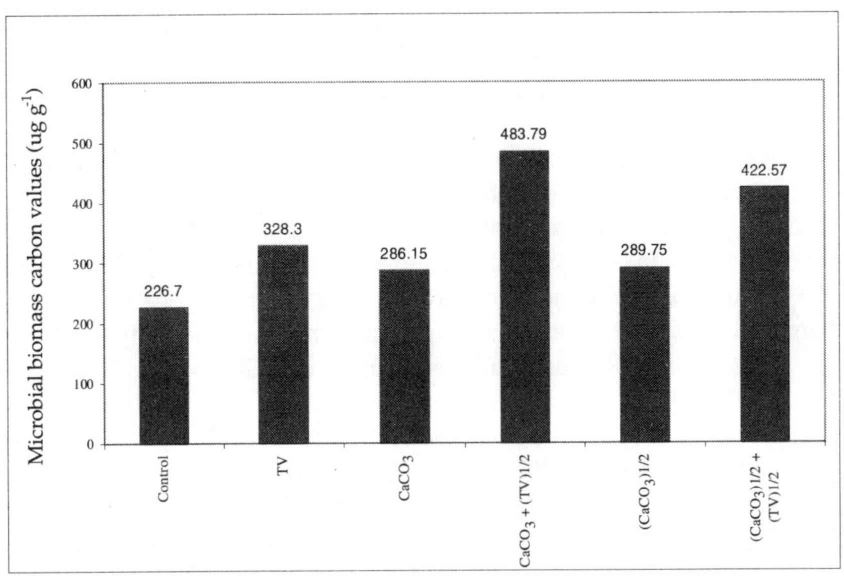

Fig-1: Microbial biomass carbon values (ug g^{-1}) under different treatments

inoculated microbes as the gut flora of these earthworms for furthering this biotechnology.

References

Bailey, V.L.; Peacock, A.D.; Smith J.L. and Bolton Jr. H. 2002: Relationship between microbial biomass determined by chloroform fumigation extraction, substrate induced respiration and phospholipids fatty acid analysis, Soil Biol. Biochem 34: 1385 - 1389.

Chattopadhyay, G.N. 2005: Vermicomposting as a biotechnological tool for recycling organic wastes. In : T.K.Ghosh, T. Chakraborty & G. Tripathi (Eds) Biotechnology in Environmental Management. Vol.1. APH Publ. Corpn, New Delhi, p. 135 - 146

Darwin, C. 1837: On the formation of vegetative mould, Trans Geol, Soc. London 2nd Series 5: 505- 509.

Edwards, C.A. and Lofty G.R. 1972: Biology of Earthworms. Chapmen and Hall London p. 285.

Franzluebbers, A.J. 1999: Introduction to symposium - microbial biomass: measurement and role in soil quality. Can J. Soil Sci., 79: 505- 506.

Gaur, A.C and Singh, Geeta 1995: Recycling rural and urban wastes through conventional and vermicomposting In: HLS Tandon (Ed.) Recycling of Crop, Animal, Human and Industrial Wastes in Agriculture. FDCO, New Delhi, p. 31 - 49.

Ismail, S.A. 1993: Vermicology - The Biology of Earthworms. Orient Longmans p. 92.

Kale R.D. 1998: Earthworms: Cinderella of Organic Farming. Prism Books Pvt. Ltd, Bangalore, PP 88.

Satchell, J.C. 1967: Nitrogen turnover by a woodland population of *Lumbricus terrestris*. In J. Doeksin and J.Vander Drift (Eds.) Soil Organisms. North Holl and Publication Amsterdam. p. 60 - 66.

Senapati, B.K. 1993: Earthworm gut contents and its significance In: Earthworms Resources and Vermiculture A.K.Ghosh (Ed), ZSI, Kolkata, p. 97 - 100.

Subba Rao, N.A. 1999: Soil Microbiology, Oxford & IBH Publ. Corpn Pvt. Ltd. New Delhi, p 407.

Tripathi, G. 2005: Earthworm - a tool for environmental biotechnology. In: T.K. Ghosh.,T. Chakraborty and G. Tripathi (Eds.) Biotechnology in Environmental Management, APH Publ. Corpn. New Delhi, p. 147 - 162.

Wallwork, J.A. 1984: Earthworm Biology. Gulab Vazirani for Arnold Heinemann, 1st Indian Edn., New Delhi.

Chapter **15**

Vermicomposting of Serifarm Waste by the Mulberry Sericultural Farmers of Orissa State

B.K.Dikshit and K.M.Purohit

Regional Sericultural Research Station, Central Silk Board, Post Box no.9, Koraput-764 020, Orissa

Orissa is sericulturally a non-traditional state where mulberry sericulture was introduced during sixth five year plan. Then with the patronage of Central Silk Board, sericultural activities were further intensified during the seventh five year plan and benefits were extended to the farmers at large (Dubey and Dikshit, 2000). At present 1246 farmers have taken up sericulture in 993 acres of land in the state. Broadly speaking in Orissa State mulberry sericulture is practiced in hilly districts with higher elevations up to 1000 meters above MSL. In these districts mostly irrigation potential is very poor as the water table is relatively low. And, the soil of these mulberry-grown areas are residual soils found in the mountains and plateaus, red in colour, acidic, poor in fertility and water holding capacity (Table-1). Majority of sericultural farmers are socio-economically very poor, small and marginal, belong to scheduled community and have half an acre of mulberry garden under rain fed conditions (Datta, 2005). Generally they prefer unutilized and dry upland for mulberry cultivation keeping whatsoever the irrigation potential land under their possession for growing other staple food crops. For each and every operation of mulberry they need financial assistance and they could not even spare any service for cultural operation of their own mulberry garden without any payment. The small size of their holdings also brought down the number of animals in their possession thus directly affecting the production of farmyard / cow dung manure for field use. Besides, cow dung is used as fuel by most of the farmers. And, whatsoever small quantity of organic manure they produce is used for growing other staple food crops as per their tradition. Hence, most of the farmers hardly apply farmyard / cow dung manure in mulberry garden. Even the farmyard / cow dung manure used by few farmers is of poor in quality. In addition to non-use of organic manure in mulberry garden, the organic matter content also depletes due to cultivation, burning of crop residues, oxidation of soil by high temperature and soil erosion etc. All these have resulted in poor fertility of soil with micronutrient deficiencies surfacing on the mulberry scene as problems (deterioration of leaf quality leading to poor health of silkworms and inferior quality of cocoons).

Table-1: Soil characteristics of different mulberry grown districts

Sl.No	Districts	Soil type	pH of soil
01	Koraput	Mostly red, mixed red & yellow; Texture: sandy loam to clay loam; Rich in Iron and Aluminium; Poor in Molybdenum, Zinc and Boron.	5.4 – 6.2
02	Rayagada	Mostly red-brown; Texture: loamy; Rich in Iron, Aluminium and Zinc; Poor in Molybdenum, Zinc and Boron.	6.0 – 7.0
03	Gajapati	Mostly red or red brown; Texture: loamy or sandy clay;	5.5 – 6.2
04	Kandhamal	Mostly red, yellow-red and black; Texture: sandy loam; Rich in Potassium; Poor in Nitrogen and Phosphorus.	6.2 – 6.5
05	Keonjhar	Mostly yellow red to grey; Texture: sandy loam; Poor in Nitrogen, Phosphorus and Potassium.	6.2 – 6.8
06	Mayurbhanj	Mostly red; Texture: sandy loam; Rich in Iron; Poor in Zinc and Boron.	5.6 - 6.8
07	Deogarh	Mostly grey to red and red-yellow; Texture: sandy loam; Medium in Nitrogen, Phosphorus and Potassium.	6.2 – 7.0

As such, vermicomposting technology was introduced and popularized among the sericultural farmers in the 10th five year plan for better productivity of leaf and cocoon by utilizing seri-farm waste to prepare vermicompost and apply the same in mulberry garden as an alternative to farmyard / cow dung manure. It is estimated that, from a half acre of mulberry garden a farmer can generate about 1.5-2 MT of seri-farm waste (weeds, grasses, unutilized leaves, silkworm litter etc.) and can use the same for vermicomposting. Vermicomposting is a simple organism-based, cost effective, affordable and user-friendly biotechnology which involves the recycling of organic matter into nutrient rich organic compost using earthworms. The earthworms feed on the decomposing organic matter to excrete the digested and partially digested matter as vermicast those put together are known as vermicompost.

Vermicomposting of seri waste by farmers

A few progressive farmers were selected for adopting vermicomposting technology and financed under Catalytic Development Project of Central Silk Board. Regional Sericultural Research Station, Koraput demonstrated Vermicomposting technology to the farmers. So far, in Koraput, Rayagada and Keonjhar districts 25 vermicomposting units were established (Table-2). Each unit comprised of a brick walled and cemented tank of size 13′ length x 4′ breadth x 3′ depth with thatched roof over the tank. The tank partitioned into two portions- one of size 7′6″ length for vermicomposting and the other of size 5′ length for preparing half decomposed seri-farm waste. The seri-farm waste (weeds, grasses, unutilized leaves, silkworm litter etc.) mixed with cow dung slurry (5 kg in 100 litres of water) was first half decomposed in the decomposing chamber for about 15 days. During decomposition the material was turned upside down. After half decomposition of the organic matter, the material was kept out of the chamber at least for two days to bring down the temperature. Then the half decomposed matter was placed in the vermicomposting chamber and earthworms *Eisenia fetida* Savigny, were inoculated into it @ 1000 numbers of earthworms / one sq mt area. The material was spread in the chamber in 3-4″ thick and maintained bed temperature of 25-30 °C and moisture of 40-60 % by sprinkling water. The material was turned upside down in every week and kept the shade almost dark, moist and cool. After 50 days, the material became black-brown coloured granular aggregates as organic manure or vermicompost. Then it is heaped in cone-shaped in middle of the chamber and allowed 2 hours to the worms for it to come down at the bottom to avoid light and for comfortable moist shelter. The compost of the upper ¾ th of the heap was spread and dried in shade and sieved through a 3 mm sieve

Table – 2: Adoption of vermicompost technology by farmers practicing sericulture of Orissa State

Year	Name of the district	Name of the block	Name of the village	No of units	Cost / Unit (Rs)	Social fabric system			
						ST	SC	OBC	Others
2004 - 2005	Koraput	Lamtaput	Bayaput	02	7,000	ST	-	-	-
			Sagar	01	-do-	ST	-	-	-
		Nandapur	Semla	02	-do-	ST	-	-	-
	Keonjhar	Ghatagaon	Dolongpani	01	-do-	ST	-	-	-
2005 - 2006	Rayagada	Kashipur	Bhagamunda	01	-do-	-	-	-	Others
			Kakudipadar	03	-do-	ST	-	-	-
			Podakhamar	10	-do-	ST	-	-	-
2006 - 2007	Koraput	Lamtaput	Jalahanjar	02	-do-	ST	-	-	-
			Sukriguda	01	-do-	-	-	-	Others
		Nandapur	Semla	02	-do-	ST	-	-	-
Total	03	04	09	25		24	Nil	Nil	02

and the small cocoons along with the tiny worms were collected. The left over worms with little quantity of vermicompost at the bottom of the heap are transferred to the fresh culture for further use. The vermicompost thus prepared were applied in mulberry field @ 5 MT / hectare / year in a single dose soon after the bottom pruning of mulberry plants in June. For comparative analysis leaf yield data control plots applied with farmyard manure were maintained. Leaf yield data of vermicompost treated and control plots were recorded separately for three crops (Rainy, Autumn and Spring seasons) in a year and analyzed (Table-3).

Table-3: Economics of vermicomposting of sericultural wastes (1 acre unit) as practiced under CDP

Sl.No.	Particulars	Quantity	Cost / Returns (Rs)
I	**Fixed costs**		
	1) Bricks	500 nos	600.00
	2) Cement	4 bags	800.00
	3) Sand		300.00
	4) Wooden poles ...10 ft height 7 ft height	3 nos 6 nos	900.00
	5) Bamboos	25 nos	500.00
	6) Thatching tiles	400 nos	2500.00
	7) Labour charges		1000.00
	8) Earthworms (starter culture)		400.00
	Total fixed costs		**7000.00**
	Depreciation 20 %		**1400.00**
II	**Variable Cost** Labour charge @ 5 mandays for 3 cycle @ Rs. 60/ manday	15 MD	900.00
	Total variable cost (Rs.1400 + 900)		2300.00
III	**Returns**		
	Income from vermicompost @ Rs. 3.00 / kg	1.5 MT	**4500.00**
IV	**Net Returns** (Rs. 4500 – 2300)		**2200.00**
V	**B : C ratio(Benefit:Cost) = 2.04:1**		

Experimental findings

It is observed that most of the farmers have half an acre of mulberry and are able to generate only 1.5 MT of seri-farm waste and prepare

700 to 800 kg of vermicompost in a year. Thus, they can meet the requirement of their mulberry garden. The leaf yield data recorded during different seasons of a year revealed that in each and every season, leaf yield of plots treated with vermicompost is somewhat higher than that of control plots applied with farmyard manure. The economics of vermicomposting was worked out (**Table** -3). The total cost for establishing one vermicompost unit under Catalytic Development Programme was estimated to Rs.7000.00, out of which 50% as subsidy and balance 50% met by the farmer. The total cost which comprises the variable cost and depreciation aggregates to Rs.2300/-. On an average a farmer has produced 700 kg of vermicompost per year and applied the same quantity in his mulberry garden @ 5 MT / ha / year. Leaf yield data was recorded for 3 crops / year and the analyzed data revealed that vermicompost treated plots registered 4 % gain over control plots applied with farmyard manure. And, with this 4 % increase, which is equivalent to 320 kg, more leaf (considering leaf yield of 8 MT/ha/year), farmers are able to rear extra 30 dfls and thereby producing 10-12 kg more cocoons.

By utilizing 3 MT of seri-waste generated from one acre mulberry, 1.5 MT of vermicompost can be prepared amounting Rs.4500/- per year. The net return amounts to Rs.2200/- per year and B: C (benefit: cost) ratio works out to 2.04:1. From these facts and figures it is inferred that vermicomposting is economically viable for mulberry sericultural farmers to recycle the Seri farm waste.

Conclusion

Vermicomposting technology adopted by the mulberry sericultural farmers proved beneficial as the farmers are able to increase the soil productivity of their mulberry garden by recycling the seri-farm waste as most of the farmers do not use farmyard manure at all. and those who are using the quality is not up to mark. The family labour available with the farmers is sufficient to take care of individual vermicomposting units, which are self-sustainable under mulberry agro system.

References

Datta, R.N.2005: Socio Economic Status Survey of Sericulture Farmers' of Orissa. RSRS, Koraput, p.1-30.

Dikshit, B.K., Purohit, K. M. and Sarkar, A. 2005: Popularization of vermiculture and effect of vermicast on mulberry leaf yield under rain fed conditions in the Eastern Ghat Highland zone of Koraput district in Orissa. Paper presented in: National Seminar on Composting & Vermicomposting, Mysore, p. 168.

Dikshit, B.K., Purohit, K.M. & Sarkar, A.2006: Status of soil health and its management practices for rain fed sericulture in Koraput District of Orissa. National Seminar on Soil Health and Water Management for Sustainable Sericulture (Abstracts), Kodathi, Bangalore, p.30-31.

Dikshit, B.K.2006: Package of practices for mulberry cultivation. RSRS, Koraput Trainers' Training Programme, p.1-14.

Dubey, A.K. & Dikshit, B.K.2000: Moriculture in Orissa. Sericulture in India. (Eds.H.O.Agrawal & M.K. Seth), Vol. p. 181-192. B. Singh & M. P. Singh, Dehradun, India.

Chapter **16**

Vermicomposting of Serifarm Wastes - An Ecofriendly Approach for Sustainable Soil Fertility Management and Rain Fed Mulberry Productivity in Koraput District of Orissa

K.M. Purohit and B.K.Dikshit
Regional Sericultural Research Station, Central Silk Board, Koraput-764 020 Orissa, India

Koraput district, situated in the southeastern region of Orissa between 18^0 10′ and 20^0 10′N latitude and 82^0 10′ and 83^0 20′ E longitude is contiguous to the mainland of eastern Ghat. It has a rolling topography with an average annual rainfall of 1522 mm. The soils are mostly red, mixed red and yellow, alluvial and red and black with sandy loam to sandy clay-loam textures, acidic and poor in fertility status, highly eroded, rich in iron and aluminium content and usually deficient in boron and zinc. They have become degraded mainly because of soil erosion resulting from excess run off, excess tillage without protection, shifting cultivation in short cycles, indiscriminate deforestation, depletion of soil nutrients and leaching of base materials, intensive cropping using high yielding varieties without organic manures, etc. Soil degradation being a major problem of the district. It is essential to use organic manure for sustainable soil fertility management and productivity.

Mulberry, the only food plant for silkworm, *Bombyx mori* L. is cultivated mainly for leaf production. Thus, the quantity of dfls reared and the cocoon yield depends on the quantity and quality of mulberry leaf produced. The recommended dose of N: P_2O_5: K_2O for irrigated mulberry in eastern India is 336:180:112 kg/ha/yr (with N in 4 split doses). alongwith 20 mt FYM/compost; while for rain fed mulberry the ratio is 150:50:50 with 10mt FYM/compost. But there is a general tendency among farmers to skip the application of FYM/compost probably because FYM/ cattle dung is in short supply due to dwindling cattle population among the farming community as also due to extensive use of cow dung cake as fuel and traditional composting is time consuming. Use of chemical fertilizers alone not only leads to pollution of ground water and eutrophication of water bodies but also adversely affects the soil fertility, micro and macro flora and fauna in the ecosystem (Upadhyaya *et al.,* 1998; Jayaraj *et al.,* 2005 and Thippeswamy *et al.,* 2005). Das *et al* (1999) and Setua *et al* (2002) *established* the superiority of vermicompost over FYM and other Seri composts and advocated its use in mulberry garden. Thus, to find out the effect of vermicompost in rain fed mulberry garden in the edapho-climatic conditions of Koraput district a study was conducted at the Regional Sericultural Research Station, Koraput farm.

What is Vermicomposting?

Vermicomposting is the technique of converting decomposable organic wastes into nutrient rich compost using selected epigeic earthworms. It is carried out in aerobic conditions for production of organic manure in shorter duration and to get rid of the foul smell of the degraded organic wastes. *Eisenia fetida* (Savigny), *Eudrilus eugeniae* (Kinberg) and *Perionyx excavatus* (Perrier) are popularly used earthworm species for this purpose. The earthworms' feed on the partially decomposed organic wastes, assimilates a negligible portion for their growth and excrete the rest in the form of nutrient rich vermicast/vermicompost.

Vermicomposting of serifarm wastes

The farm residues such as left over mulberry leaves, soft twigs and farm weeds as also silkworm litter and rearing wastes offer a good opportunity to generate nutrient rich and ecofriendly vermicompost. For this purpose an existing cemented pit of 7′6″x4′6″x3′ dimension with tin roof covered with thatch is used as vermi-pit. Another open pit of 8′x4′x3′ dimension is used for preparation of semi-decomposed material. First of all, the farm residues were treated with cow dung solution and allowed for partial decomposition for 15 days in the open pit. Then the partially decomposed material was loaded into the vermi-pit to form a bed of 9″ depth at a desired moisture content of 50-60%. After checking the temperature in bed the earthworm species of *Eisenia fetida* (3000 number) were released into the bed. The materials were turned carefully by a forked spade at periodic intervals and an ideal environmental condition was maintained with respect to moisture 50-60%, temperature 25-30^{0}C and a pH level almost to neutral. For maintaining the desired level of moisture a measured quantity of water was applied by sprinkling. To maintain the desired temperature in the bed certain insulating material like gunny bag, paddy straw etc were used. The worms fed on the partially decomposed material and excreted the faecal matter in the form of cast. After about six weeks the material, appeared soft, spongy and dark brown in colour with earthy odor (no foul smell). Then the vermicast was dumped in a conical heap and left for a few hours. The worms from the base of the conical heap were collected and re-used. Earthworms were separated by screening through a net of 2 mm size. After earthworm separation the moisture content of vermicompost was brought down to 20-25% (by drying) before bagging. The vermicompost, thus produced, was applied @ 5 mt/ha/yr in June in the established mulberry garden having S_{1635} mulberry genotype with a spacing of 90x90 cm and compared with FYM (@ 10 mt/ha/yr) applied plot. Both the plots were maintained following the recommended package of practices for rain fed mulberry and three leaf harvests were made every year.

Experimental findings

Leaf yield data recorded over the two years indicated that application of 5 mt vermicompost/ha/yr along with N: P_2O5 : K_2O @ 150 : 50 : 50 kg /ha/yr produced 6.8 % more leaf *vis-à-vis* the use of 10 mt FYM/ha/yr along with the same quantity of N : P_2O_5 : K_2O (Table-1) . This may be due to the fact that vermicompost contained higher amount of NPK compared to FYM/conventional compost (Das *et al.*, 1997;

Table - 1. Leaf yield (kg/ha/yr) as influenced by vermicompost application

Treatments	2004-05	2005-06	Average	% of increase over control
FYM @10 mt/ha/yr (control)	6744	7425	7084	--
Vermicompost @ 5 mt/ha/yr	7384	7755	7569	6.8

Dandin *et al.*, 2000; Sarkar, 2005) (**Table-2**), as also due to the enhanced efficiency of fertilizers added and increased fertility of soil due to organic manure application (Laxminarayan and Patiram, 2005). It is reported that vermicompost contained certain active ingredients similar to that of hormones and mulberry cuttings planted after treating with vermicompost slurry recorded significantly higher number of roots and higher survival rate of cuttings (Doriglan and Gowda, 1996; Balikai *et al.*, 1998). Presence of such substance in the vermicompost might have contributed to root proliferation in mulberry leading to increased nutrient uptake resulting in good growth and leaf yield. It is also reported that application of vermicompost in soil improved the leaf quality of S_{1635} mulberry genotype in terms of increased protein and sugar contents compared to the leaves produced by using FYM/ conventional compost albeit bioassay study indicated no difference (Setua *et al*, 2002; Sarkar, 2005).

Table 2. NPK contents of different organic manures (Das *et al.*, 1997; Dandin *et al.*, 2000)

Organic manures	Nutrient Content (%)		
	N	P	K
Vermicompost	1.87	0.6	1.0
Compost	0.6	0.7	0.3
FYM	0.3	0.15	0.5

Conclusion

It may be concluded that seri farm wastes have the greater potentiality to be used for preparing vermicompost, which can be effectively used to meet organic manure requirements for the sericulture farmers.

Acknowledgement

The authors express their sincere thanks to the Joint Director, Regional Sericultural Research Station, Central Silk Board, Koraput, for providing necessary facilities and encouragement.

References

Balikai, R.A., Rajashekar, D.W. and Patil, G.M. 1998: Influence of FYM, vermicompost, RDF, Mycorrhiza and their combinations on the establishment of mulberry cuttings in summer nursery. *Advances in Agricultural Research in India*. **9**: 147 – 150.

Dandin, S. B., Jayaswal, J. and Giridhar, K. 2000: Handbook of sericulture technologies. Central Silk Board, Bangalore.

Das, P.K., Bhogesha, K., Sundareswaran, P., Madhava Rao Y. R. and Sharma, D. D. 1997: Vermiculture: scope and potentiality in sericulture. *Indian Silk*, **36(2)**: 23- 26.

Das, P.K., Theertha Prasad B. M., Bhogesha, K., Katiyar, R. S., Vijayakumari, K. M. and Madhava Rao Y. R. 1999: Effect of graded doses of vermicompost of sericultural wastes on mulberry growth, leaf yield and quality. In: Moriculture in tropics, vol. I. *Proc. Natl. Seminar on Tropical Sericulture*, Dept of Sericulture, Univ. Agric. Sci, GKVK, Bangalore and Seri 2000. Swiss Agency for Development and Co-operation, Bangalore. p. 28-30.

Doriglan, S.B.and Gowda, R.R.1996: Effect of vermicompost on rooting of mulberry cuttings.*Proc.Natl. Seminar on organic farming and sustainable sericulture*. UAS, Bangalore. 45-46.

Jayaraj, S., Veeraiah, T. M., Qadri, S. M. H., Amarnath, S., Jaishankar Srinivas Rao T. V. S., Rao Ramamohana, Samuthiravel P., Ravikumar, J., Chandrasekhar Reddy D. and Dandin, S.B. 2005: Studies on the impact of integrated nutrient management on mulberry yield, soil health and economics in three states in south India through farmer participatory mode. *20th Congr. Internatl. Seric, Commission.Central Silk Board, Bangalore Conference Papers*. Vol.**1**, sec.1.Mulberry p. 68-76.

Laxminarayana, K. and Patiram. 2005: Influence of inorganic, biological and organic manures on yield and nutrient uptake of ground nut (*Arachis hypogea*) and soil properties. *Ind. J. Agric. Sci.* **75(4):** 218 - 220.

Sarkar, A. 2005: Vermicomposting of seri farm residues at CSR&TI, Berhampore and its impact on leaf production in mulberry in eastern India. *National Seminar on Composting and Vermicomposting* (Lead papers and Abstracts), CSR&TI, Mysore. P. 93-97.

Setua, G. C., Banerjee, N.D., Sengupta, T, Das, N. K., Ghosh, J.K. and Saratchandra, B. 2002: Comparative efficacy of different composts in S_1 mulberry *Morus alba* L. under rain fed condition. *Indian J. Agric. Sci.* **72 (7)**: 389 - 392.

Thippeswamy, T., Das, P.K., Subbaswamym M.R. and Chandrakantm K.S., 2005: Integrated technology package for sustenance of mulberry leaf yield and cocoon production. *20th Congr. Internatl. Seric. Commission.Central Silk Board, Bangalore Conference Papers*. Vol.**1**, sec.1.Mulberry p. 37 - 40.

Upadhyaya, A.K., Srinivasa Reddy, D.V. and Biddappa, C.C. 1998: Organic farming technology for coconut. *Indian Coconut J.* **24:** 74-78.

Chapter **17**

Effect of Mono and Polycultures Earthworms on Recycling of Non-Mulberry Seri wastes under Hygrothermic Conditions

K.V.Shankar Rao and G.P. Mohobia

Regional Sericultural Research Station, Landiguda, Koraput-764020, Orissa

**Regional Tasar Research Station, Dharampura, Jagdalpur-494 005, Chhattisgarh.*

Approximately 4000 species of earthworms have been identified of which 500 species are reported from India of these only about 20 species have the potential for the use in vermicomposting and even fewer three to five species can be handled and cultured in organic waste and are most widely used in the sericulture based vermicomposting process. In Chhattisgarh State, About 44% of the total area (135.1 sq. km.) is under forest cover and occupies primary position in Vanya Silk production. Tasar silkworm rearing are conducted in 29540 acres (Table-1) and obtains about 200 MT of Seri waste which can be recycled and converted into 120 MT. of vermicompost and can generate additional income resource for poor tribal of the state. Thus there is urgent need to suggest a viable earthworm species or mixed culture for optimisation of the biological process. Hence a study was undertaken to ascertain the efficacy of earthworm mono and polyculture under Chhattisgarh hygrothermic conditions.

TABLE - 1. District-wise farming indicators of Tasar Culture Development in Chhattisgarh

Sl. No.	Name of the district	TASAR			
		Arjuna Economic Plantation (acres)	Rearing forest area (acres)	Total rearing area (acres)	No. of beneficiaries
01	Jangir	873.75	522.75	1396.50	374
02	Bilaspur	476.25	1296.75	1773.00	200
03	Korba	1611.16	4420.02	6031.18	2013
04	Ambikapur	1171.25	1912.50	3083.75	382
05	Korea	521.25	540.00	1061.25	81
06	Raigarh	2935.00	4929.88	7864.88	3977
07	Jashpur	1690.00	1096.50	2786.50	513
08	Mahasamund	37.75	205.50	243.25	957
09	Dhamtari	40.00	75.00	115.00	26
10	Kawardha	-	-	-	-
11	Durg	-	-	-	-
12	Rajanandgaon	55.00	50.00	105.00	185
13	Raipur	48.12	37.50	85.62	147
14	Jagdalpur	257.77	951.50	1209.27	185
15	Dantewara	490.00	1850.25	2340.25	223
16	Kanker	254.82	1190.25	1445.07	304
	Total	**10462.12**	**190078.40**	**29540.52**	**9579**

Source: Directorate of Rural Industries (Sericulture Sector), Govt. of Chhattisgarh, Raipur

Vermicomposting of seri waste (Polyculture method)

A vermery consisting of four trenches each measuring 10'x05'x1.5' was constructed. Seri waste biomass of 400 kg. was mixed with 200 kg. of cow dung and allowed to decompose for 20 days. Thus each trench was filled with 600 kg. of semi-decomposed material and 6000 numbers of *E. eugeniae, E. fetida* and *P. excavatus* were released in to three trenches separately. In the fourth trench 2000 earthworms of *E. eugeniae, E. fetida* and *P. excavatus* of each were released (polyculture). All the trenches were protected from direct sunlight, heat and rain by raising thatched shed using local materials. The shed was meant to provide shade and to keep ambient atmosphere cool and humid. A small channel was also dug around the shed to drain excess rainwater during monsoon to avoid water logging. Earthworm activity was monitored and observations were recorded and analysed for composting duration, quantum of vermicompost harvested and rate of conversion for mono and polyculture to ascertain the efficiency. The cycle was repeated for five times in a year.

Experimental findings

The composting period is ranged from 41-50 days for pre composted seri-material (Table-2) and the conversion rate ranged from 63.33-76.77 % for different mono and polyculture. However, the duration was less for polyculture treatment followed by *E. fetida, E. eugeniae* and *P. excavatus*. The reason for this may be attributed to combined effect of three earthworm species. The conversion of wastes in to vermicompost ranged from 63.33 to 76.67% of the raw materials in different mono and polyculture. However maximum conversion of waste was observed in polyculture consisting of *E. eugeniae* + *E. fetida* + *P.excavatus*. This may be due to the combined effort of the mixed culture consisting of three earthworm species active in one season or other.

TABLE - 2 Conversion rate of vanya silk sericultural residues with different earthworm species and their poly-culture

Treatment (Earthworm species)	Quantum of Seri Organic Waste (Kg.)	Composting Duration (Days)	Quantity of Vermi compost Harvested (Kg)	Rate of Conversion
E. Eugeniae (Kinberg)	600	48	400	66.67
E. fetida (Savigny)	600	45	420	70.00
P. excavatus (Perrier)	600	50	380	63.33
Polyculture (Ee + Ef + Pe)	600	41	460	76.67
	SD % CV%	$\pm$ 1.732 6.580	$\pm$ 34.641 29.428	- -

Perionyx excavatus distributed in India, *E. eugeniae* known as African right crawler and *E.fetida* originating from Caucasus region are commonly used in vermicomposting process even in poly-cultures (Reddy, 2005). Species of earthworms viz. *E. eugeniae, E. fetida and P. excavatus* have been the effective breeds of worms for waste management (Kale *et al.*, 1982; Kale and Bano, 1988; Giraddi and Lingappa; 2000 and Reinecke *et al.*, 1992). The Present study confirms that utilization of mixed culture (poly culture) comprising of three earthworm species is effective for vermicomposting process. The present investigation concludes that polyculture (*E. eugeniae* +*E. fetida*

+ *P.excavatus*) > *E. fetida* > *E. eugeniae* > *P. excavatus* in the conversion rate of seri waste in to vermicompost.

Conclusion

It is recommended to use mixed (polyculture) earthworm culture for bioconversion of tasar Seri waste material.

References

Giraddi, R.S. and Lingappa, S. 2002: Population growth rate of three detrivore species of earthworms in field designs used for bioconversion of organic wastes. Proceedings 6th International Symposium on Earthworm Ecology, 1-6 September, Cardiff University, Wales (U.K.).

Kale, R.D., Bano. K. and Krishnamoorthy, R.V. 1982: Potential of *Perionyx excavatus* for utilizing organic wastes. *Pedobiologia,* 23:419-425.

Kale, R.D. and Bano, K. 1988: Earthworm cultivation and culturing techniques for production of vermicompost, Mysore J. Agric. Sci., 22:339-344.

Reinecke, A.J., Vilioen, S.A. and Saayman, R.J, 1992: The suitability of *Eudrilus eugeniae, Perionyx excavatus* and *Eisenia fetida* (Oligochaeta) for vermicomposting in southern Africa in terms of their temperature requirements. Soil Biol. and Biochem. 24:1295-1307.

Reddy,M.V. 2005: Vermicomposting using different species of earthworms: A comparative Account. National Seminar on composting and vermicomposting 26th, 27th October. CSR & Ti, Mysore. p: 22 -25.

Chapter **18**

Economic Model of Vermicompost Production Technology using Mulberry Seri waste

R. Sahu, K.V. Shankar Rao and K.C. Brahma
Regional Sericultural Research Station, Landiguda, Koraput-764 020(Orissa)

Mulberry Sericulture is considered to be a cash crop of par excellence in the traditional states like Karnataka, Andhra Pradesh and Tamil Nadu. These areas are practicing sericulture since generations and familiar with mulberry cultivation and silkworm rearing. Adoption of modern technologies alongwith well developed marketing system enabled this culture as one of the best cash crop after grape cultivation. The agro-climatic condition of Orissa, though congenial for sericultural activities under rain fed conditions by the small and marginal tribal farmers, it is more or less subsistence farming and due to non-availability of competitive marketing system being non-traditional area is one of the major constraint. To enhance income level of the farmers for sustainability of the mulberry sericulture, there is a need to be looked into. Hence, the farm residues of sericulture are generally considered as a waste and one hectare of mulberry farm generates 10 MT of seri waste, which could be utilized for production of vermicompost through suitable low cost technology for an additional income to boost their socio-economic condition.

Economic model of vermicompost production technology

Central Sericultural Research and Training Institute, Berhampore, West Bengal evolved a economic model of vermicompost production technology which includes two pits of the size 12′ x 3′ x 3′, two polythene sheets to cover the mud walled tanks and a thatched roof for protection from rains and high temperature. The total expenditure was about Rs.2300/- per unit. It will accommodate 6 MT of partially decomposed organic food materials. Farm yard waste and seri waste were mixed with cow dung in the ratio of 8: 1 and put in the pit. This feed is allowed to decompose for at least 2 - 3 weeks. The partially decomposed biomass in the pit inoculated with 2000 earthworms, *Eisenia fetida* as these rapidly converted seri waste into casts and also breed very fast. The worms feed on the biomass assimilating 5 - 10 % for their growth and excreting the rest in the form of nutrient rich vermicompost. The entire material was converted in to vermicompost within 4 to 5 weeks time. The methodology has been demonstrated to 24 farmers of Koraput and Rayagada districts and the economics has been derived in the form of Benefit - Cost ratio.

Experimental findings

The economics of the economic model of vermicompost technology using 2000 earthworms in the 1st year has been shown in table-1. A perusal of the data indicated that out of total 12 acres of mulberry plantations, a quantity of 54.07 MT of seri waste was generated from which 42.17 MT of vermicompost was produced @ 1.76 MT of vermicompost / farmer / half acre / year. Hence, each farmer realized Rs.3520/- by selling 1.76 MT of vermicompost and thus, obtained a net profit of Rs.1220/- in the first year which gives about 2: 1 Benefit - Cost Ratio and thus become viable in the initiating year and the life span of the pit was five years, roof and polythene sheet was two years. Hence, in the second year a farmer could realize Rs.3520/- as there is no expenditure involved and 3rd to 5th year the expenditure was the bare minimum. The study also revealed that average amount realized from the sale of commercial cocoon was estimated as Rs.5000/- per half acre per annum. Thus, out of the total income of Rs.6220/-, the vermicompost has generated about 20 % of additional income and this has acted as incentive to the farmers and increased the sustainability of mulberry sericulture in the tribal districts of Orissa. Thus, the preliminary evaluation of the economic model indicates that the activity can be taken up on a commercial scale. Further with an active participation of Government agencies, NGOs and farmers themselves compost and vermicompost technologies are needed to be popularized extensively to increase adoption levels by the farmers and to further augment the production and use of vermicompost.

Table - 1 Performance of Economic Model of Vermicompost Production Technology in Orissa

Area of mulberry garden		12.0 acre
No of beneficiaries		24 tribal farmers
Total seri waste generated		54.07 MT
Total production of vermicompost		42.17 MT
Production of vermicompost / farmer / year		1.76 MT
A. Expenditure :		
1 Construction of 2 Nos. thatched roof / polythene Sheet covered mud walled tank / chamber (size : 12' x 3' x 3') for approximately 6 MT partially decomposed food materials.		Rs.1000.00
2. Cost of initial requirement of @ Rs.15/- per 100 earthworms.		Rs. 300.00
3. Labour (20 MDs) @ Rs40/ M.D		Rs. 800.00
4. Miscellaneous		Rs. 200.00
TOTAL:		**Rs. 2300.00**
B. Income :		
Sale proceeds of 1.76 MT vermicompost produced during the 1st year @ Rs.2/- per kg.		= **Rs.3520.00**
C. **Net Profit :**	Rs.3520/- - Rs.2300/-	= **Rs.1220.00**
D. **Benefit Cost Ratio:**		= **2.00: 1.00**

Index